SOLUTIONS TO PROBLEMS IN ELECTRONIC COMMUNICATIONS SYSTEMS

OKO ANYAEGBU
BSC; MTECH; MIET; C ENG; COREN

outskirts
press

Solution to Problems in Electronic Communications Systems
All Rights Reserved.
Copyright © 2019 Oko Anyaegbu BSc; MTech; MIET; C Eng; COREN
V2.0

The opinions expressed in this manuscript are solely the opinions of the author and do not represent the opinions or thoughts of the publisher. The author has represented and warranted full ownership and/or legal right to publish all the materials in this book.

This book may not be reproduced, transmitted, or stored in whole or in part by any means, including graphic, electronic, or mechanical without the express written consent of the publisher except in the case of brief quotations embodied in critical articles and reviews.

Outskirts Press, Inc.
http://www.outskirtspress.com

ISBN: 978-1-9772-0250-5

Cover Photo © 2019 www.gettyimages.com. All rights reserved - used with permission.

Outskirts Press and the "OP" logo are trademarks belonging to Outskirts Press, Inc.

PRINTED IN THE UNITED STATES OF AMERICA

Foreword

Engr. Anyaegbu has produced an excellent introduction for both undergraduate and post graduate students in Telecommunication Engineering to the dominant means of transmission in the Telecommunication Network, viz:

Wire pairs and coaxial cables.
Microwave transmission system and satellite communication.
Fiber optic transmission.

The theory and solutions to problems will serve as a good introduction especially in the practical areas of wireless traffic backhaul design. While microwave has been the dominant transmission system, the demand for greater bandwidth with the advent of 3G and lately 4G has necessitated the use of fiber cables.
I highly recommend this book for final year undergraduate and masters students.

Prof. Monima S. Briggs.
B.Sc. (Eng.)(Imperial College), M. S. E. (Princeton),
M. E.(M.I.T.), Ph.D. (Imperial College.),
A. C. G. I., D. I. C. MNSE,
Director.
Centre for Information and Telecommunication Engineering,
University of Port Harcourt.

Preface

This book originated from lecture notes and textbooks used in teaching Communication Systems to final year undergraduate students in the University of Port Harcourt, Nigeria.

Each chapter consists of the following three sections:
- Some theoretical work.
- A number of worked examples of typical problems.
- Problems with answers for the students to practice.

The theoretical work consists of brief notes dealing with some aspects of the subject only. No attempt has been made to produce a complete textbook. However, it is hoped that the notes will be useful in the solution of the problems.

Dedication

This book is dedicated to God who made it possible. It is also dedicated to my wonderful wife, Elder (Mrs.) Ezi O. Anyaegbu, our wonderful children: Dr. (Mrs.) Esther O. A. Ihekweazu, Dr. (Mrs.) Elizabeth I. A. Onugha, Dr Margaret U. A. Dike, Grace N. Anyaegbu, Anyaegbu O. Anyaegbu; our wonderful sons – in law, Mr Andrew I. Onugha, Engr Okechukwu Ihekweazu and Mr Churchill C. Dike and our wonderful daughter – in – law, Mrs. Onyinye Anyaegbu. You all encouraged and supported me in the writing of this book. May God continue to bless you all in Jesus Name – Amen.

Symbols and Abbreviations

C	Capacitance
c	Velocity of light
D	Diameter
d	Diameter; Distance
E	Electric Field Strength
F	Force
f	Frequency
G	Conductance; Universal Gravitation Constant
g	Normalized Conductance
H	Magnetic Field Strength
h	Height
I	Current
j	Vector operator ($\sqrt{-1}$)
k	Voltage Reflection Coefficient
k_r	Solution of Bessel Function Equation
L	Inductance
l	Length
M	Mass
m	Number of half wavelengths across a waveguide of width a

n	Refractive Index; Number of half wavelengths across a waveguide of height b.
P	Power
P_F	Power Flow
p	Propagation constant per unit length of a uniform line
R	Resistance
R_0	Characteristic Resistance of a line
s	Standing Wave Ratio
T	Period
t	Time
V	Voltage
v	Velocity
w	Angular Frequency ($2\pi f$)
X	Reactance
Y	Admittance
Y_0	Characteristic Admittance
y_L	Normalized Load Admittance
Z	Impedance
Z_0	Characteristic Impedance of a line
z_L	Normalized Load Impedance

α	Attenuation Constant per unit length of a uniform line; Angle
β	Phase Constant
δ	Increment
ε	Permittivity
ε_0	Permittivity of free space
ε_r	Dielectric constant of material
ϕ	Angle; Phase Angle
λ	Wavelength
μ	Permeability
μ_0	Permeability of free space
π	Ratio of circumference to diameter of circle
θ	Angle; Phase Angle
θ_c	Critical Angle
ρ	Standing wave ratio
τ	Time Delay
$\Delta\tau$	Modal Dispersion

EHF	Extremely High Frequency
GEO	Geostationary Orbit
HF	High Frequency
INTELSAT	International Telecommunication Satellite Organization
LEO	Low Earth Orbit
LF	Low Frequency
MEO	Medium Earth Orbit
MF	Medium Frequency
NA	Numerical Aperture
OC	Open Circuit
SC	Short Circuit
SHF	Super High Frequency
TE	Transverse Electric Mode
TM	Transverse Magnetic Mode
UHF	Ultra High Frequency
VHF	Very High Frequency
VLF	Very Low Frequency

Table of Contents

Foreword i
Preface iii
Dedication v
Symbols and Abbreviations vii
Table of Contents xi
List of Figures xv
List of Tables xix
1 Transmission Media 1
   1.1 Introduction 1
2 Transmission Lines 3
   2.1 Introduction 3
   2.2 Parallel Wire 3
   2.3 Twisted Pair 4
   2.4 Coaxial Cable 4
   2.5 Characteristic Impedance 5
   2.6 Equivalent circuit of a transmission line 7
   2.7 Reflection from an imperfect termination 9
   2.8 The Smith chart and its applications 12
   2.9 Impedance Matching in Transmission Lines 15
         2.9.1 Quarter Wavelength Line 16
         2.9.2 Tapered Transmission Line 17
         2.9.3 Stub Matching 18
   2.10 Further Examples on Transmission Lines 20
   2.11 Questions on Transmission Lines 37
3 Fiber-Optics 40
   3.1 Introduction 40
   3.2 Applications of Fiber-Optics 40
         3.2.1 Road Signs 41
         3.2.2 Endoscopes 41
         3.2.3 Lighting Hazardous Areas 41

3.2.4 Lighting Ships at Sea 41
3.2.5 Telecommunications 42
3.3 Light Wave Spectrum 42
3.4 Refraction... 43
3.5 Critical Angle 44
3.6 Snell's Law 45
3.7 Reflection... 46
3.8 Problem of Power Loss... 47
3.9 Solution to the Problem of Power Loss caused by dirt................ 48
3.10 Propagation of light along the fiber 49
3.11 Numerical Aperture 50
3.12 Cone of Acceptance 52
3.13 Modes in Fiber Optics... 52
3.14 Multimode Step Index Fiber 54
3.15 Problem of Modal Dispersion.. 55
3.16 Solution to Modal Dispersion 58
    3.16.1 Multimode Graded Index Fiber 58
    3.16.2 Single Mode Step Index Fiber 59
3.17 Cut-off Wavelength 60
3.18 Cable Composition.. 61
3.19 Fiber Sizes.. 61
3.20 Light Sources and Detectors for Optical Cables........................ 62
    3.20.1 Injection Laser Diode.. 62
    3.20.2 Light Emitting Diodes (LEDs)... 63
    3.20.3 PIN Diodes... 63
    3.20.4 Avalanche Photo Diode 64
3.21 Advantages of Optical Fiber 64
3.22 Disadvantages of Optical Fiber.. 65
3.23 Connecting Optical Fibers 65
    3.23.1 Compatibility 66
    3.23.2 Gap Loss 69
    3.23.3 Alignment Problems 70
3.24 Splicing of Optical Fibers 70
    3.24.1 Fusion Splicing 71
    3.24.2 Mechanical Splicing.. 71
3.25 Examples on Fiber Optics 71
3.26 Questions on Fiber Optics.. 80

4 Waveguides 83
4.1 Introduction 83
4.2 Applications 84
4.3 Advantages of Waveguides 85
4.4 Method of Wave Propagation in a Waveguide 85
4.5 Dominant Mode of Propagation 88
4.6 Plane Waves at a Conducting Surface 88
4.7 Phase Velocity 90
4.8 Cut-off Wavelength 90
4.9 Cut-off Frequency 92
4.10 Group and Phase Velocities 93
4.11 Impedance Concept 95
4.12 Methods of Exciting Waveguides 96
4.13 Circular Waveguides 97
4.14 Differences in behaviour between circular and rectangular waveguides 98
4.15 Advantages of circular waveguides 98
4.16 Disadvantage of circular waveguides 99
4.17 Examples on Waveguides 99
4.18 Questions on Waveguides 116

5 Antennas 119
5.1 Definition 119
5.2 Radiation from a current element in free space 119
5.3 Power radiated by a doublet 122
5.4 Antenna Losses and Efficiency 124
5.5 Antenna Gain 126
5.6 Gain of a Doublet 126
5.7 Polar Diagram 127
5.8 Power radiated by a short dipole 129
5.9 Power radiated by a half-wavelength dipole in free space 130
5.10 Power radiated by a short vertical earthed antenna 131
5.11 Effect of the earth on antennas 133
5.12 General theory of antenna array 136
5.13 Broadside Array 139
5.14 Polar Diagram of Broadside Array 140
5.15 Beam Angle of Broadside Array 141
5.16 End-fire Array 142

5.17 Beam angle of the end-fire array 144
5.18 Rhombic Antenna 146
5.19 Yagi-Uda Antenna 148
5.20 Antennas with parabolic reflectors 149
5.21 Examples on Antennas 151
5.22 Questions on Antennas 167
6 Unguided Media (Wireless) 169
   6.1 Introduction 169
   6.2 Frequency Allocation 169
   6.3 Types of Propagation 170
      6.3.1 Surface Propagation 172
      6.3.2 Tropospheric Propagation 173
      6.3.3 Ionospheric Propagation 173
      6.3.4 Line-of-sight Propagation 174
      6.3.5 Space Propagation (Satellite Communication) 175
   6.4 Kepler's Three Laws of Planetary Motion 177
      6.4.1 The law of ellipses 177
      6.4.2 The law of equal areas 177
      6.4.3 The law of harmonies 177
      6.4.4 A demonstration of the third law 178
   6.5 Fundamentals of satellite communication 180
      6.5.1 Advantages of satellite communications. [6] 181
      6.5.2 Disadvantages of satellite communications. [6] 183
   6.6 Satellite orbits 183
      6.6.1 Geostationary orbit (GEO) 183
      6.6.2 Low/ Medium earth orbit (LEO/ MEO) 184
   6.7 Historical overview of satellite communication. [6] 185
   6.8 Examples on Satellite Communications 187
   6.9 Questions on Satellite Communication 194
References 196

List of Figures

Fig 1.1:	Classification of transmission media	1
Fig 2.1:	Parallel wire	3
Fig 2.2:	Twisted pair	4
Fig 2.3:	Coaxial cable	4
Fig 2.4:	Transmission Line Geometry	6
Fig 2.5:	Uniform Transmission Line	7
Fig 2.6:	Equivalent Circuit of a Transmission Line	7
Fig 2.7:	Smith chart solution of Example 1	13
Fig 2.8:	Stub matching	19
Fig 2.9:	Reflection at the point of connection of two transmission lines	23
Fig 2.10:	Position of voltage minimum on a line	28
Fig 2.11:	Smith chart solution of Example 6	32
Fig 2.12:	Smith chart solution of Example 7	35
Fig 3.1:	Light wave spectrum	42
Fig 3.2:	Light from less dense medium to denser medium	43
Fig 3.3:	Light from denser medium to less dense medium	44
Fig 3.4:	Effect of increasing the angle of incidence	45
Fig 3.5:	Refraction at a junction of two media	45

Fig 3.6:	Reflection in fiber	47
Fig 3.7:	Power loss due to contamination	48
Fig 3.8:	Core surrounded with cladding	48
Fig 3.9:	Light spreading at the end of the fiber	49
Fig 3.10:	Total internal reflection	50
Fig 3.11:	Cone of acceptance	52
Fig 3.12:	Multimode step index fiber	54
Fig 3.13:	Modal dispersion in fiber optics	55
Fig 3.14:	Multimode graded index fiber	58
Fig 3.15:	Single mode step index fiber	59
Fig 3.16:	Fiber construction	61
Fig 3.17:	Large core to small core	66
Fig 3.18:	Small core to large core	66
Fig 3.19:	Losses due to unequal numerical aperture	68
Fig 3.20:	Gap Loss	69
Fig 3.21:	Loss due to lateral misalignment	70
Fig 4.1:	a) Rectangular and b) Circular waveguides	83
Fig 4.2:	Transverse electromagnetic wave transmission	86
Fig 4.3:	Wave propagation in a waveguide	86
Fig 4.4:	Reflection from a conducting surface	87
Fig 4.5:	Wavefronts incident on a perfectly conducting plane-reflections are not shown	88

Fig 4.6:	Reflection in a parallel plane waveguide	90
Fig 4.7:	Loop and probe coupling of waveguides	96
Fig 5.1:	Radiation from a current element in free space – elementary doublet (Hertzian dipole)	120
Fig 5.2:	Total power radiated by a doublet	123
Fig 5.3:	Polar diagram of an elementary doublet	128
Fig 5.4:	Current distribution in a short dipole	129
Fig 5.5:	Current and voltage distributions in a half-wavelength dipole	130
Fig 5.6:	Vertical earthed antenna	132
Fig 5.7:	Direct and reflected rays from an antenna	134
Fig 5.8:	Antenna arrays with fields	137
Fig 5.9:	Radiation pattern of broadside array	140
Fig 5.10:	Beam angle of broadside array	141
Fig 5.11:	Polar diagram for the end-fire array	143
Fig 5.12:	Beam angle of the end-fire array	144
Fig 5.13:	Rhombic antenna	146
Fig 5.14:	Rhombic antenna radiation pattern	147
Fig 5.15:	Yagi-Uda antenna and radiation pattern	148
Fig 5.16:	a) Geometry of a parabola b) Antenna with parabolic reflector	150
Fig 5.17:	Field due to two antennas in end-fire array	151

Fig 5.18:	Service area of the antennas in end-fire array	153
Fig 5.19:	Field due to two antennas in broadside array	155
Fig 5.20:	Service area of the antennas in broadside array	156
Fig 5.21:	Field due to array of 4 antennas	158
Fig 5.22:	Polar diagram of an array of 4 antennas	163
Fig 5.23:	Polar diagram of two element end-fire array	166
Fig 6.1:	The electromagnetic spectrum. [4]	169
Fig 6.2:	The radio communications spectrum. [4]	170
Fig 6.3:	The Earth with surrounding layers of atmosphere	171
Fig 6.4:	Surface Propagation	172
Fig 6.5:	Tropospheric Propagation	173
Fig 6.6:	Ionospheric Propagation	174
Fig 6.7:	Line-of-sight Propagation	175
Fig 6.8:	Satellite Communication. [4]	176
Fig 6.9:	A planet in near circular orbit around the sun	178
Fig 6.10:	A satellite communication system	181
Fig 6.11:	Comparison of satellite and terrestrial communication costs	182
Fig 6.12:	Figure for Example 1	188
Fig 6.13:	Figure for Example 2	190
Fig 6.14:	Figure for Example 3	193

List of Tables

Table 3.1:	Fiber Types	61
Table 3.2:	Losses due to unequal core sizes. [3]	67
Table 3.3:	Losses due to unequal numerical aperture. [3]	68
Table 4.1:	Examples of rectangular waveguides	84
Table 4.2:	Values of k_r for the principal modes in circular waveguides [1]	97
Table 5.1:	Values of radiation field in different directions	128

1
Transmission Media

1.1 Introduction

Electrical energy is not always used where it is produced and means have to be provided to transmit the energy from one point to another by some form of electromagnetic wave propagation. Electromagnetic signals can travel through a vacuum, air or other transmission media.

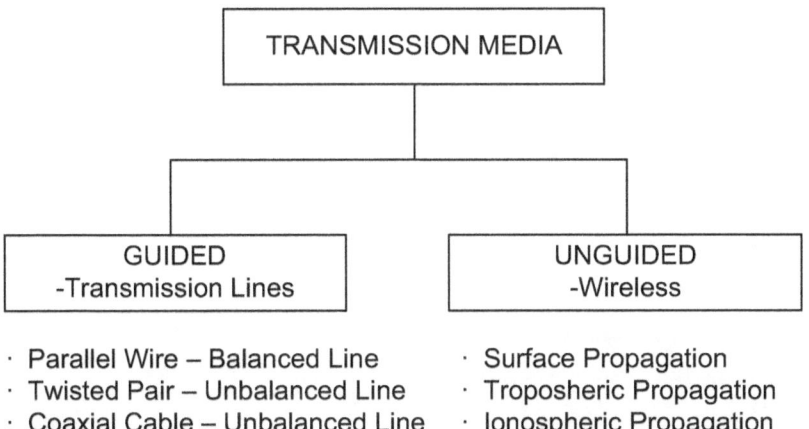

Fig. 1.1: Classification of transmission media

Fig. 1.1 shows the two main methods by which energy can be transmitted from one point to another. The first method employs guides (Transmission Lines). In this case, electromagnetic waves are guided along a solid medium such as twisted pair, coaxial cable, waveguide, or optical fiber. The second method is unguided. It employs radiation of free electromagnetic waves and is usually referred to as wireless transmission. Unguided transmission is achieved by using antennas.

2
Transmission Lines

2.1 Introduction

Transmission lines are impedance-matching circuits designed to deliver radio frequency power from the transmitter to the antenna and maximum signal from the antenna to the receiver [1].

2.2 Parallel Wire

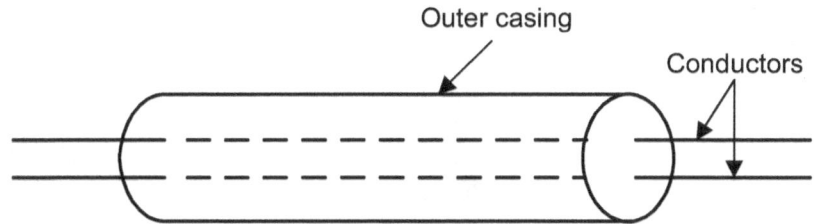

Fig. 2.1: Parallel wire

The parallel wire is illustrated in Fig. 2.1. It is used where balanced properties are required, such as when connecting a folded dipole antenna to a TV receiver. It has a problem of electromagnetic interference from devices such as electric motors. This creates noise. The parallel wire is not used at microwave frequencies.

2.3 Twisted Pair

A twisted pair cable usually has between 6 – 40 twists per meter, as depicted in Fig. 2.2. This reduces noise. The upper frequency limit is 5MHz.

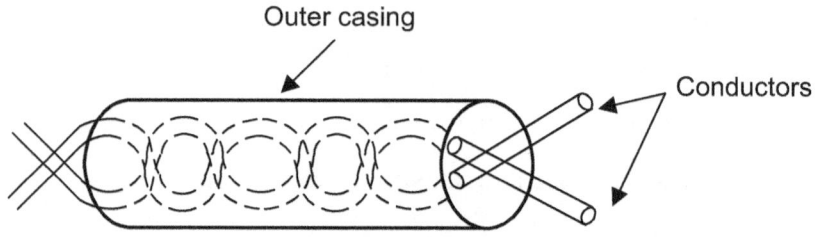

Fig. 2.2: Twisted pair

2.4 Coaxial Cable

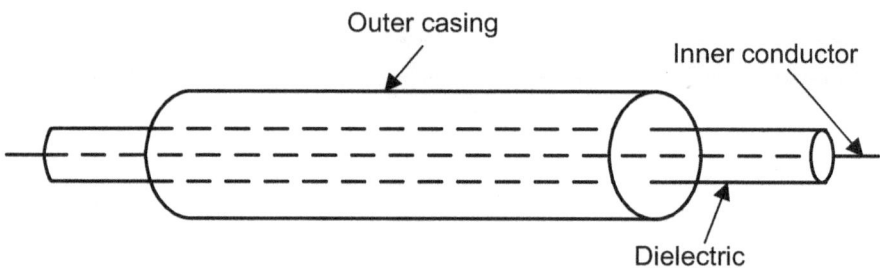

Fig. 2.3: Coaxial cable

The coaxial line is used when unbalanced properties are needed, such as when connecting a broadcast transmitter to its grounded antenna. It is used at frequencies up to 18 GHz. Coaxial lines may be rigid or flexible, air-spaced or filled with different dielectrics.

The cross-sectional dimension of the cable determines the maximum power capability. For example, a rigid air-dielectric coaxial copper cable with outer diameter of 22.5 mm has a peak power rating of 43 kW. This increases to 400 kW for an outer diameter of 80 mm, and to 3 MW for a 230 mm outer diameter. However, as the outer diameter is increased the inner diameter must also be increased, in order to ensure a constant value of all the line properties [1].

2.5 Characteristic Impedance

The characteristic impedance of a transmission line, Z_0, is the impedance measured at the input of the line when its length is infinite. The characteristic impedance depends on the geometry of the transmission line. Fig. 2.4 illustrates the transmission line geometry for a parallel wire and a coaxial cable.

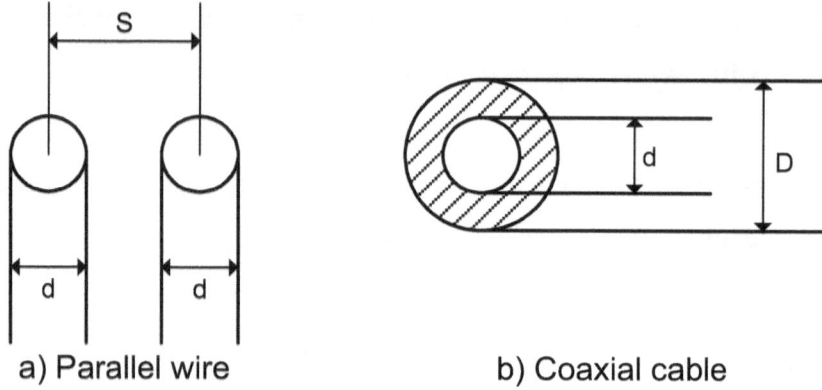

Fig. 2.4: Transmission Line Geometry

For the parallel wire, the characteristic impedance is given by:

$$Z_0 = \frac{276}{\sqrt{\varepsilon_r}} \log\left(\frac{2s}{d}\right) \text{ ohms}, \quad \ldots\ldots 2.1)$$

where ε_r is the dielectric constant of the insulation, d is the diameter of the dialectric surrounding each conductor, and s is the distance between the two parallel conductors.

For the coaxial line, the characteristic impedance is given by:

$$Z_0 = \frac{138}{\sqrt{\varepsilon_r}} \log\left(\frac{D}{d}\right) \text{ ohms}, \quad \ldots\ldots 2.2)$$

where ε_r is the dielectric constant of the insulation, d is the diameter of the inner conductor, and D is the diameter of the outer conductor.

The usual range of Z_0 is 150 – 600 ohms for balanced lines (parallel wire or twisted pair) and 40 – 150 ohms for coaxial lines. [1]

2.6 Equivalent circuit of a transmission line

A uniform transmission line is shown in Fig. 2.5, while its equivalent circuit is illustrated in Fig. 2.6.

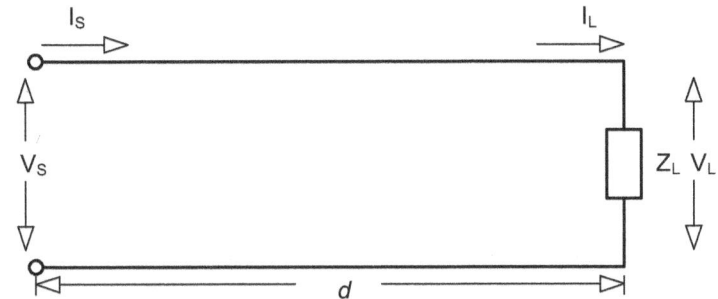

Fig. 2.5: Uniform Transmission Line

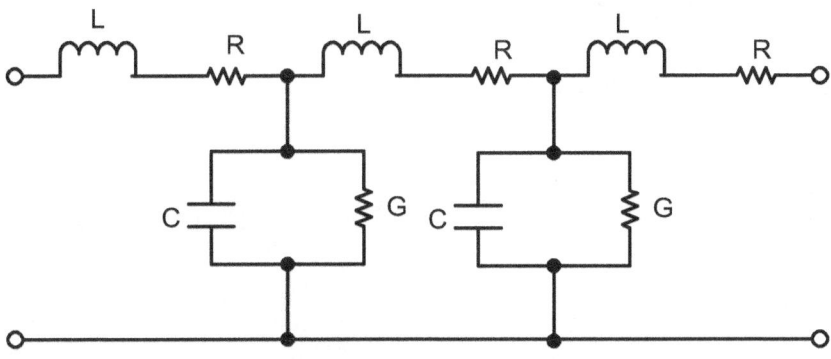

Fig. 2.6: Equivalent Circuit of a Transmission Line

The four parameters or primary constants of a transmission line, shown in Fig. 2.6, are as follows:

L = Loop inductance per unit length;
R = Loop resistance per unit length;

G = Shunt conductance per unit length; and

C = Shunt capacitance per unit length.

If the applied voltage V_s in Fig. 2.5 is sinusoidal, then it can be shown that:

$$V_L = V_s \cosh \rho l - (I_s Z_0) \sinh \rho l \quad \text{............(2.3)}$$

$$I_L = I_s \cosh \rho l - \left(\frac{V_s}{Z_0}\right) \sinh \rho l, \quad \text{............(2.4)}$$

where V_s, I_s, V_L, and I_L are RMS voltages and currents.

ρ is the propagation constant per unit length which is given by:

$$\rho = \sqrt{(R + jwL)(G + jwC)} \quad \text{............(2.5)}$$

The characteristic impedance of the line, Z_0, is:

$$Z_0 = \sqrt{\frac{R + jwL}{G + jwC}} \quad \text{............(2.6)}$$

The input impedance is given by:

$$Z_s = Z_0 \left[\frac{Z_L \cosh \rho l + Z_0 \sinh \rho l}{Z_0 \cosh \rho l + Z_L \sinh \rho l}\right] \quad \text{............(2.7)}$$

The propagation constant per unit length, ρ, can also be represented as:

$$\rho = \alpha + j\beta, \quad \text{............(2.8)}$$

where α is the attenuation constant in nepers per unit length, and β is the phase constant in radians per unit length.

If λ is the wavelength of the line, then:

$$\lambda = \frac{2\pi}{\beta} \quad\quad\quad\quad\quad\quad\quad\quad\quad\quad\quad\quad\quad\quad\quad\quad\quad\quad (2.9)$$

If the phase velocity is v_p, then:

$$v_p = \lambda f = \frac{2\pi f}{\beta} \quad\quad\quad\quad\quad\quad\quad\quad\quad\quad\quad\quad\quad\quad\quad (2.10)$$

The constants ρ, α, β, and Z_0 are known as the secondary line constants.

2.7 Reflection from an imperfect termination

If a loss-free transmission line has infinite length or is terminated in its characteristic impedance, all the power applied to the line by the generator at the input will be absorbed by the load. If a finite piece of line is terminated in an impedance not equal to the characteristic impedance, then some of the applied power will be absorbed by the load while the remaining power will be reflected.

The voltage reflection coefficient is defined as:

$$k = \frac{Z_L - Z_0}{Z_L + Z_0}, \quad\quad\quad\quad\quad\quad\quad\quad\quad\quad\quad\quad\quad\quad\quad\quad\quad\quad (2.11)$$

where Z_L is the load impedance and Z_0 is the characteristic impedance.

The voltage transmission coefficient is given by:

$$\frac{2Z_L}{Z_L + Z_0} \quad\quad\quad\quad\quad\quad\quad\quad\quad\quad\quad\quad\quad\quad\quad\quad\quad\quad\quad\quad (2.12)$$

The current reflection coefficient is given by:

$$\frac{Z_0 - Z_L}{Z_L + Z_0} = -k \quad\quad\quad\quad\quad\quad\quad\quad\quad\quad\quad\quad\quad\quad\quad\quad\quad\quad (2.13)$$

The current transmission coefficient is given by:

$$\frac{2Z_0}{Z_L + Z_0} \quad\quad\quad\quad\quad\quad\quad\quad\quad\quad\quad\quad\quad\quad\quad\quad\quad\quad\quad\quad (2.14)$$

The voltage standing wave ratio is defined as:

$$S = \frac{|V_{max}|}{|V_{min}|}, \quad\quad\quad\quad\quad\quad\quad\quad\quad\quad\quad\quad\quad\quad\quad\quad\quad\quad\quad (2.15)$$

where V_{max} is the maximum voltage along the transmission line and V_{min} is the minimum voltage across the transmission line.

$$S = \frac{1+|k|}{1-|k|} \quad \text{...............(2.16)}$$

Also,

$$|k| = \frac{S-1}{S+1} \quad \text{...............(2.17)}$$

At high frequencies transmission lines are considered as loss-free.

From Equation (2.8), $\rho = \alpha + j\beta$.

If the line is loss-free, the attenuation $\alpha = 0$.

Therefore, $\rho = j\beta$.

From Equation (2.7), $Z_s = Z_0 \left[\dfrac{Z_L \cosh \rho l + Z_0 \sinh \rho l}{Z_0 \cosh \rho l + Z_L \sinh \rho l} \right]$.

If $\rho = j\beta$, then:

$$Z_s = Z_0 \left[\frac{Z_L \cos \beta l + jZ_0 \sin \beta l}{Z_0 \cos \beta l + jZ_L \sin \beta l} \right] \quad \text{...............(2.18)}$$

From Equation (2.6), the characteristic impedance is given as:

$$Z_0 = \sqrt{\frac{R + jwL}{G + jwC}}$$

At radio frequencies, $R \ll wL$ and $G \ll wC$.

Hence,

$$Z_0 = \sqrt{\frac{jwL}{jwC}} = \sqrt{\frac{L}{C}}, \quad \text{...............(2.19)}$$

where L is measured in Henrys/meter and C is measured in Farads/meter.

From Equation (2.19), it is clear that the characteristic impedance is resistive at radio frequencies.

2.8 The Smith chart and its applications

The Smith chart, devised by P. H. Smith, is an impedance chart which consists of constant resistance and reactance circles or arcs of circles plotted on a polar diagram. The diagrams are so arranged that various important quantities connected with mismatched transmission lines may be plotted and evaluated fairly easily. An example is shown in Fig. 2.7.

Some of the applications of the Smith chart will be illustrated with the following example.

Example 1

A line of characteristic impedance $Z_0 = 50\,\text{ohm}$ is terminated in a load of $50 + j50\,\text{ohm}$. With the aid of a Smith chart, calculate:
 a) The magnitude and phase of the reflection coefficient,
 b) The standing wave ratio,
 c) The position of voltage maximum,
 d) The position of voltage minimum, and
 e) The admittance of the load.

Solution

Refer to Fig. 2.7.

$Z_L = 50 + j50$ ohms

$Z_0 = 50$ ohm

∴ The normalised load impedance is given by:

$$z_L = \frac{Z_L}{Z_0}$$

$$= \frac{50 + j50}{50}$$

$$= 1 + j$$

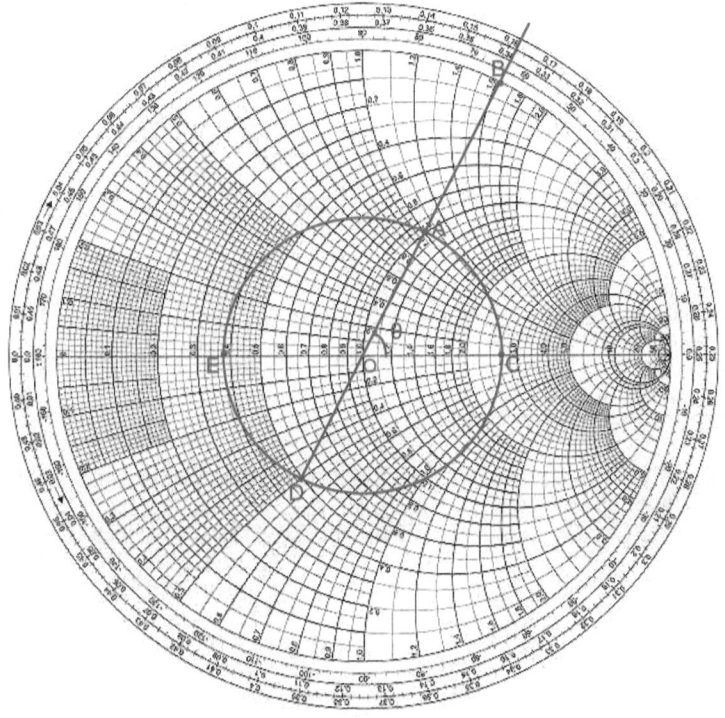

Fig. 2.7: Smith chart solution of Example 1

On the Smith chart in Fig. 2.7, locate the normalised load at point A. With the centre at point O and radius OA draw a circle. This circle represents the load. It is the circle of constant standing wave ratio. Draw the line OA and extend it to the circle of radius $r = 0$ at point B.

a) The reflection coefficient is found by the length of OA as a fraction of the radius of the unity circle $r = 0$.

$$\text{Reflection coefficient} = \frac{OA}{OB} = \frac{35}{80} = 0.438$$

The phase angle of the reflection coefficient is $\theta = 64° = 1.12$ radians.

The reflection coefficient can also be expressed as:

$k = 0.438\angle 64°$
$k = 0.438(\cos 64° + j\sin 64°)$
$k = 0.19 + j0.39$

b) The standing wave ratio is the intersection of the circle of constant standing wave ratio and the horizontal straight line corresponding to zero reactance and to the right of the centre of the circle of constant standing wave ratio. This is read off the chart as 2.6.

c) From the load at point A, move clockwise towards the generator until you come to point C which represents the point of pure resistance. This is read off the chart circumference as 0.089λ.

C represents the point of maximum voltage. The position of the voltage maximum can also be obtained from the angle $\theta = 64°$.

On the Smith chart, $180°$ is equivalent to 0.25λ.

Hence, $\theta = 64° = \dfrac{0.25\lambda}{180} \times 64 = 0.089\lambda$.

d) From the load move clockwise towards the generator until you come to point E which represents the point of pure resistance. This is equivalent to $0.089\lambda + 0.25\lambda = 0.339\lambda$.

E represents the point of voltage minimum.

e) Project the line BAO until it meets the circle of constant standing wave ratio at point D.

Point D represents the normalized admittance of the load.

Point D is read off the chart as $0.5 - j0.5$.

Hence $y_L = 0.5 - j0.5$.

But $Z_0 = 50$ ohms.

Hence $Y_0 = \dfrac{1}{50} = 0.2$.

The load admittance is given by:

$Y_L = y_L \times Y_0 = 0.2(0.5 - j0.5)$

$Y_L = 0.01 - j0.01 S$.

2.9 Impedance Matching in Transmission Lines

If the load connected to a transmission line has an impedance which is equal to the characteristic impedance of the line, then the line is said to be matched. In this condition the reflection coefficient $k = 0$ and so the standing wave ratio is 1.0. Most systems are therefore designed to work with the standing wave ratio as near to 1.0 as possible. A value of standing wave ratio greater than 1.0 represents mismatch and leads to loss of power at the receiving end. In other cases, it may cause a voltage breakdown as in high power radar systems or distortion as in televisions [2]. It is therefore important to be able to match a line. At high frequencies the following methods are used to match transmission lines: quarter wavelength method, tapering, and stub-matching.

2.9.1 Quarter Wavelength Line

As given in Equation (2.18), the general expression for the input impedance of a loss-free line of length l is given by:

$$Z_s = Z_0 \left[\frac{Z_L \cos \beta l + jZ_0 \sin \beta l}{Z_0 \cos \beta l + jZ_L \sin \beta l} \right].$$

When $l = \dfrac{\lambda}{4}$ or an odd number of quarter-wavelengths, then we have:

$$\beta l = \beta \frac{\lambda}{4}.$$

But $\beta = \dfrac{2\pi}{\lambda}$.

Hence $\beta l = \dfrac{2\pi}{\lambda} \times \dfrac{\lambda}{4} = \dfrac{\pi}{2}.$

$$\therefore Z_s = Z_0 \left[\frac{Z_L \cos\frac{\pi}{2} + jZ_0 \sin\frac{\pi}{2}}{Z_0 \cos\frac{\pi}{2} + jZ_L \sin\frac{\pi}{2}} \right]$$

$$= \frac{Z_0^2}{Z_L} \quad \dotfill (2.20)$$

This relationship is called Reflective Impedance.

If Z_0 is varied, the impedance seen at the input to the quarter wavelength transformer will be varied accordingly and the load can be matched to the characteristic impedance of the main line. This is similar to varying the turns ratio of a transformer to obtain a required value of input impedance for any given value of load impedance. The quarter wavelength line is frequency dependent.

2.9.2 Tapered Transmission Line

If the spacing of the conductors of a transmission line is varied, the impedance of the line varies. A line in which the spacing varies uniformly from the input provides a gradual impedance transformation along its length. Impedance matching can be achieved by interposing a suitably designed tapered line between the load and the main transmission line. The advantage of a tapered line is that it is less frequency selective than a quarter wavelength line.

Recall from Equation (2.1) that the charactersitic impedance of a parallel wire is given by:

$$Z_0 = \frac{276}{\sqrt{\varepsilon_r}} \log\left(\frac{2s}{d}\right) \text{ ohms,}$$

where ε_r is the dielectric constant of the insulation, d is the diameter of the dialectric surrounding each conductor, and s is the distance or spacing between the two parallel conductors. If s varies, then Z_0 varies.

2.9.3 Stub Matching

If a short section of loss-free line is connected across a main transmission line it will introduce a reactance at the point of attachment. The value of the reactance will depend on the length of the section and on the condition of the end remote from the main line. Such sections of line are referred to as stubs and are used for impedance matching. As shown in Fig. 2.8, the stub is connected in parallel with the main line. Hence for calculations on stub-matching, it is preferable to deal with admittances rather than impedances.

A loss-free stub of variable length should be considered as a device which has zero input conductance and a susceptance which varies with the stub length. If the terminal load on a transmission line is not equal to the characteristic impedance then the input admittance of the line will vary with the distance from the load. The stub is placed at a point on the line where the input conductance at that point is equal to the characteristic conductance of the line. The stub length is adjusted to provide a susceptance which is equal in value but opposite in sign to the input susceptance of the main line at that point. Then the total susceptance at the point of the stub attachment will be zero. The combination of stub and line will thus present a conductance which is equal to the characteristic conductance of the line. Hence the main length of high frequency line will be matched. If the line is several wavelengths long there will be a multiplicity of points at which the stub may be placed. However, it is normally positioned at the point nearest the load. This ensures that the greatest possible length of the line is operated in a matched condition. Stubs may have the end remote from the main line open-circuited or short-circuited. The short-circuited stub is generally preferred in practice because open-circuited pieces of transmission line tend to radiate from the open end [1].

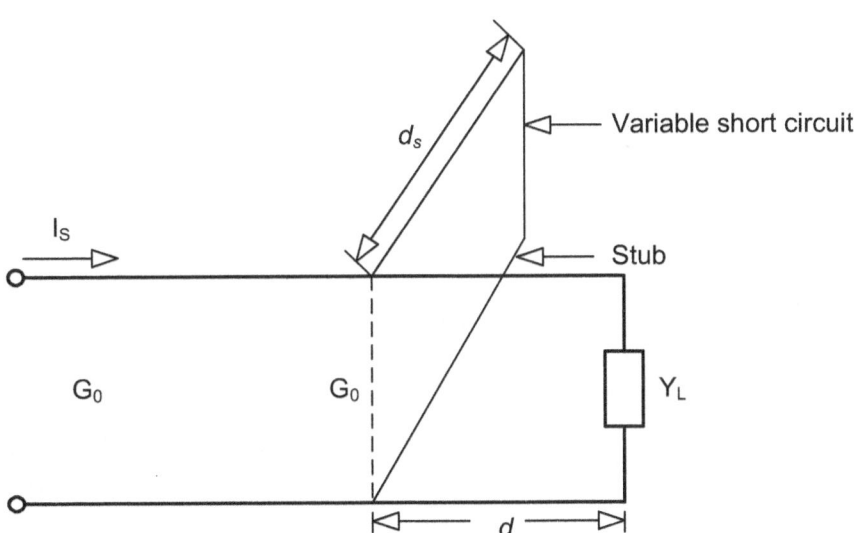

Fig. 2.8: Stub matching

2.10 Further Examples on Transmission Lines

Example 2

A 50 ohm line is terminated in a load impedance of $70 + j60$ ohms. The line is 2m long and is excited by a source of energy at 50MHz. Assume that the velocity of propagation along the line is equal to the speed of light in free space. Calculate:

    a) The input impedance
    b) The magnitude and phase of the voltage reflection coefficient,
    c) The voltage standing wave ratio,

Solution

$Z_L = 70 + j60$ ohms

$Z_0 = 50$ ohms

$$\lambda = \frac{c}{f} = \frac{3 \times 10^8}{50 \times 10^6} = 6 \text{ m}$$

$$\beta = \frac{2\pi}{\lambda} = \frac{2\pi}{6}$$

$l = 2$ m

$$\therefore \beta l = \frac{2\pi}{6} \times 2 = 120°.$$

Transmission Lines

a) $Z_s = Z_0 \left[\dfrac{Z_L \cos \beta l + j Z_0 \sin \beta l}{Z_0 \cos \beta l + j Z_L \sin \beta l} \right]$

$= 50 \left[\dfrac{(70 + j60)\cos 120° + j50 \sin 120°}{50 \cos 120° + j(70 + j60)\sin 120°} \right]$

$= 50 \left[\dfrac{(70 + j60)(-0.5) + j50(0.866)}{50(-0.5) + j(70 + j60)(0.866)} \right]$

$= 50 \left[\dfrac{-35 - j30 + j43.3}{-25 + j60.62 - 51.96} \right]$

$= 50 \left[\dfrac{-35 + j13.3}{-76.96 + j60.62} \right]$

$= 50 \left[\dfrac{37.44 \angle -20.8°}{97.97 \angle -38.2°} \right]$

$= 19.11 \angle 17.4°$

$= 18.24 + j5.71 \text{ ohms}.$

b) The voltage reflection coefficient is given by:

$k = \dfrac{Z_L - Z_0}{Z_L + Z_0}$

$= \dfrac{70 + j60 - 50}{70 + j60 + 50}$

$= \dfrac{20 + j60}{120 + j60}$

$= \dfrac{63.25 \angle 71.6°}{134.16 \angle 22.6°}$

$= 0.47 \angle 45°$

$= 0.33 + j0.33$

c) The voltage standing wave ratio is given by:

$$S = \frac{1+|k|}{1-|k|}$$

$$= \frac{1+0.47}{1-0.47}$$

$$= \frac{1.47}{0.53}$$

$$= 2.77$$

Example 3

a) Define the characteristic impedance of a transmission line.

b) A transmission line of characteristic impedance Z_0 is connected to a second line of characteristic impedance Z_R which is correctly terminated. By considering the reflection at the junction of the two lines, derive an expression for the voltage reflection coefficient in terms of the characteristic impedances of the two lines.

c) A transmission line with a characteristic impedance of 150 ohms is terminated in a purely resistive load. It is found, by measurement that the minimum value of voltage on the line is 4 micro volts and the maximum is 10 micro volts. Calculate the value of the load resistance.

Solution

a) The characteristic impedance of a transmission line is defined as the impedance at the input terminals of an infinite length of the line.

b) Fig. 2.9 shows the reflection at the point of connection of two transmission lines.

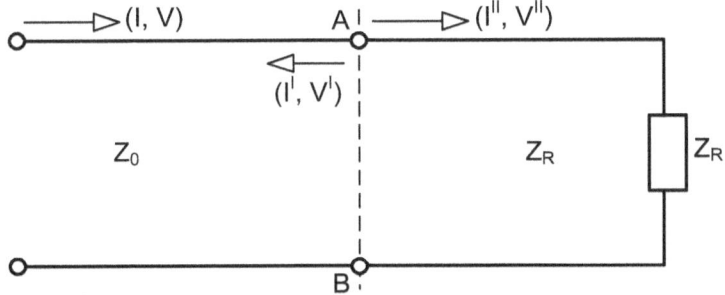

Fig. 2.9: Reflection at the point of connection of two transmission lines

(I, V) = Current, voltage of incident wave

(I', V') = Current, voltage of reflected wave

(I'', V'') = Current, voltage of transmitted wave

From Fig. 2.9, we have:

$$I'' = I + I' \quad \text{.................(2.21)}$$

$$V'' = V + V' \quad \text{.................(2.22)}$$

Also,

$$I = \frac{V}{Z_o}$$

$$I'' = \frac{V''}{Z_R}$$

$$I' = \frac{-V'}{Z_o}$$

From Equation (2.21), we have:

$$\frac{V''}{Z_R} = \frac{V}{Z_o} - \frac{V'}{Z_o}$$

$$V'' = \frac{Z_R}{Z_0}(V - V')$$

From Equation (2.22), we have:

$$V'' = V + V'$$

$$\therefore \frac{Z_R}{Z_0}(V - V') = V + V'$$

$$Z_R(V - V') = Z_0(V + V')$$

$$(Z_R - Z_0)V = V'(Z_R + Z_0)$$

$$\therefore \frac{V'}{V} = \frac{(Z_R - Z_0)}{(Z_R + Z_0)}$$

Hence the voltage reflection coefficient is given by:

$$\frac{V'}{V} = \frac{Z_R - Z_0}{Z_R + Z_0}$$

c) The standing wave ratio is given by:

$$S = \frac{|V_{max}|}{|V_{min}|}$$

$$= \frac{V + V'}{V - V'}$$

$$\therefore S = \frac{1+\dfrac{V'}{V}}{1-\dfrac{V'}{V}}$$

But $\dfrac{V'}{V} = \dfrac{Z_R - Z_0}{Z_R + Z_0}$

$$\therefore S = \frac{1+\dfrac{Z_R - Z_0}{Z_R + Z_0}}{1-\dfrac{Z_R - Z_0}{Z_R + Z_0}}$$

$$= \frac{Z_R + Z_0 + Z_R - Z_0}{Z_R + Z_0 - Z_R + Z_0}$$

$$= \frac{2Z_R}{2Z_0}$$

$$= \frac{Z_R}{Z_0}$$

Hence, $Z_R = SZ_0$

$$= Z_0 \frac{|V_{max}|}{|V_{min}|}$$

$Z_0 = 150$ ohms

$V_{max} = 10$ microvolts

$V_{min} = 4$ microvolts

$\therefore Z_R = 150 \times \dfrac{10}{4} = 375$ ohms

Example 4

Bridge measurement on a length of transmission line at 800Hz gave the following results:

Input impedance with the load short-circuited (Z_{SC}) = Resistance of 198 ohms in series with a capacitance of $5.7 \times 10^{-6} F$, Input impedance with the load open-circuited (Z_{OC}) = Resistance of 103 ohms in series with a capacitance of $1.6 \times 10^{-6} F$. Calculate:

a) The characteristic impedance of the line.
b) The four primary line constants, given that the propagation constant per km of the line at the same frequency is $0.21 + j0.24$

Solution

a) $Xc_1 = \dfrac{1}{2\pi fC_1}$

$= \dfrac{1}{2\pi \times 800 \times 5.7 \times 10^{-6}}$

$= 34.9$ ohms

$Xc_2 = \dfrac{1}{2\pi fC_2}$

$= \dfrac{1}{2\pi \times 800 \times 1.6 \times 10^{-6}}$

$= 124.3$ ohms

$\therefore Z_{SC} = 198 - j34.9$ ohms

$$= 201.1\angle -10.0° \text{ ohms}$$

$$Z_{OC} = 103 - j124.3 \text{ ohms}$$

$$= 161.4\angle -50.4° \text{ ohms}$$

$$Z_O = \sqrt{Z_{SC} \times Z_{OC}}$$

$$= \sqrt{201.1\angle -10.0° \times 161.4\angle -50.4°}$$

$$= 180.2\angle -30.2°$$

$$= 155.7 - j90.6 \text{ ohms}$$

b) $\rho = 0.21 + j0.24 /\text{km}$

$$= 0.319\angle 48.8° /\text{km}$$

$$\rho = \sqrt{(R + jwL)(G + jwC)}$$

$$Z_0 = \sqrt{\frac{R + jwL}{G + jwC}}$$

$$\rho \times Z_0 = R + jwL$$

$$\therefore R + jwL = 0.319\angle 48.8° \times 180.2\angle -30.2°$$

$$= 57.48\angle 18.6°$$

$$= 54.5 + j18.33$$

$\therefore R = 54.5 \text{ ohm/km}$

$2\pi \times 800 L = 18.33$

$$\therefore L = \frac{18.33}{2\pi \times 800}$$

$$= 3.6 \text{ mH/km}$$

$$\frac{\rho}{Z_0} = G + jwC$$

$$\therefore G + jwC = \frac{0.319\angle 48.8°}{180.2\angle -30.2°}$$

$$= 0.00177\angle 79°$$

$$= 3.38\times 10^{-4} + j1.737\times 10^{-3}$$

$$\therefore G = 0.34\times 10^{-3} \text{ mhos/km}$$

$$2\pi \times 800 C = 1.737\times 10^{-3}$$

$$\therefore C = \frac{1.737\times 10^{-3}}{2\pi \times 800}$$

$$= 0.346\times 10^{-6} \text{ F/km}$$

Example 5

The standing wave ratio on an ideal 70-ohms line is measured as 3.2 and a voltage minimum is observed 0.23 wavelength in front of the load. Find the load impedance.

Solution

Standing wave ratio $S = 3.2$

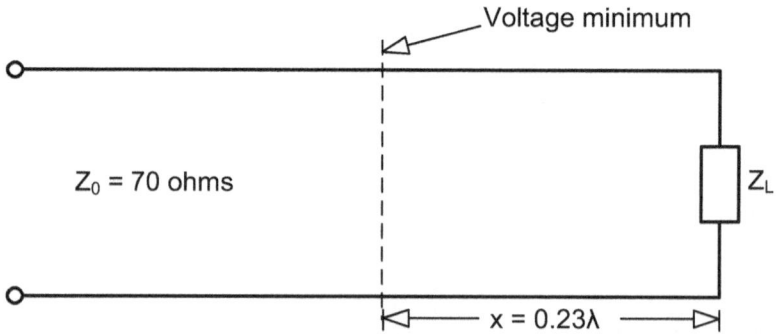

Fig. 2.10: Position of voltage minimum on a line

Let the voltage reflection coefficient at the load be given as $|k|e^{j\phi}$.

But $S = \dfrac{1+|k|}{1-|k|}$

$\therefore |k| = \dfrac{S-1}{S+1}$

$= \dfrac{3.2-1}{3.2+1} = \dfrac{2.2}{4.2} = 0.524$

$\therefore k = 0.524e^{j\phi}$

Let the voltage reflection coefficient at the voltage minimum be k_x.

$\therefore k_x = 0.524e^{j(\phi-2\beta x)}$

But $x = 0.23\lambda$

$\beta = \dfrac{2\pi}{\lambda}$

$\therefore \beta x = \dfrac{2\pi}{\lambda} \times 0.23\lambda$

$= 0.46\pi$

Hence $k_x = 0.524e^{j(\phi-0.92\pi)}$.

At the voltage minimum the incident and reflected waves are in anti-phase. Hence we have:

$\phi - 0.92\pi = -\pi$

$\phi = -\pi + 0.92\pi$

$= -0.08\pi$

$= -0.251 \text{ radians}$

Hence the reflection coefficient at the load is given by:

$$k = 0.524e^{-j0.251}$$
$$= 0.524\angle -14.38°$$
$$= 0.51 - j0.13$$

$$k = \frac{Z_L - Z_0}{Z_L + Z_0}$$

$$k(Z_L + Z_0) = Z_L - Z_0$$
$$kZ_L + kZ_0 = Z_L - Z_0$$
$$Z_0(k+1) = Z_L(1-k)$$
$$\therefore Z_L = Z_0 \frac{(k+1)}{(1-k)}$$

$$= 70\frac{(0.51 - j0.13 + 1)}{(1 - 0.51 + j0.13)}$$

$$= 70\frac{(1.51 - j0.13)}{(0.49 + j0.13)}$$

$$= 70\frac{1.516\angle -4.92°}{0.507\angle 14.86°}$$

$$= 209.31\angle -19.78°$$

$$= 197 - j70.83 \text{ ohms}$$

Example 6

A 50 ohm line is terminated in a load impedance of $75 - j70$ ohm. The line is 3.5 meters long and is excited by a source of energy at 50 MHz. Assume that the velocity of propagation along the line is 3×10^8 m/s. Use a Smith chart to calculate the following:

    a) The input impedance

b) The magnitude and phase of the voltage reflection coefficient,

c) The voltage standing wave ratio,

d) The position of voltage minimum.

Solution

Frequency of operation, $f = 50\,\text{MHz}$

Velocity of propagation, $c = 3 \times 10^8\,\text{m/s}$

But $c = \lambda f$, where λ is the wavelength.

$$\therefore \lambda = \frac{c}{f} = \frac{3 \times 10^8}{50 \times 10^6} = 6\,\text{m}$$

Length of line, $l = 3.5\,\text{m}$

$$= \frac{3.5}{6}$$
$$= 0.583\lambda$$

Load impedance, $Z_L = 75 - j70$ ohms

Characteristic impedance, $Z_0 = 50$ ohm

∴ The normalised load impedance is given by:

$$z_L = \frac{Z_L}{Z_0}$$
$$= \frac{75 - j70}{50}$$
$$= 1.5 - j1.4$$

On the Smith chart in Fig. 2.11, locate the normalised load impedance at point A. With the centre at point O and radius OA draw a circle. This circle represents the load. It is the circle of constant voltage standing wave ratio.

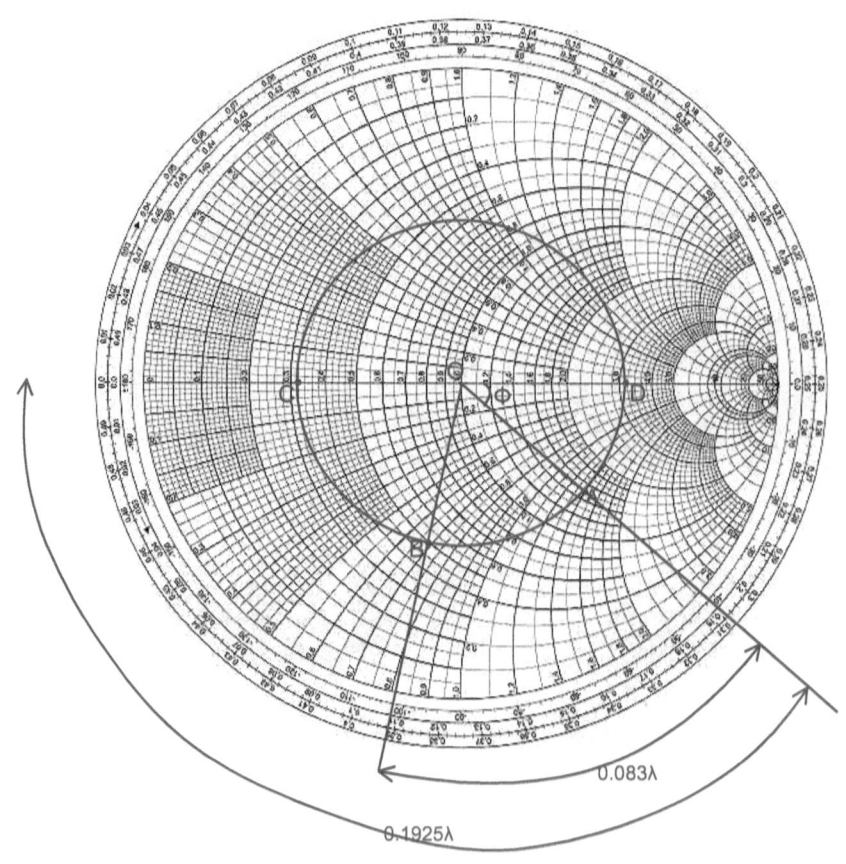

Fig. 2.11: Smith chart solution of Example 6

a) From point A, move 0.583λ towards the generator in order to locate the input. Note that the complete circumference of the Smith Chart is equivalent to 0.5λ. Hence from point A move 0.083λ towards the generator to locate the input at point B.

The normalized input impedance is read off the chart as:

$z_i = 0.5 - j0.7$

$Z_0 = 50\,\text{ohm}$

$Z_i = z_i \times Z_0$

$\quad = 50(0.5 - j0.7)$

$\quad = 25 - j35\,\text{ohm}$

b) The reflection coefficient is represented by the line $|OA|\angle\phi$.

$$\text{Reflection coefficient, } k = \frac{4.15}{8}\angle -41.5°$$

$\quad = 0.52\angle -41.5°$

$\quad = 0.52e^{-j0.72}$

c) The standing wave ratio is read off the chart at point D as 3.18.

d) The point of voltage minimum is represented by point C. This is given by the distance from the load A to the point C of minimum voltage. This is equal to 0.9125λ.

But $\lambda = 6\,\text{m}$.

Hence the position of voltage minimum from the load is $0.1925 \times 6 = 1.155\,\text{m}$.

Example 7

A certain transmission line is terminated in a load impedance of $240 - j180$ ohm. If the line has a characteristic impedance of 300 ohms determine, using the Smith Chart,:

    a) The voltage standing wave ratio,

    b) The magnitude and phase of the voltage reflection coefficient,

    c) The value, position and length of a short-circuited stub that will provide a perfect match between the line and the load,

Solution

Load impedance, $Z_L = 240 - j180$ ohms

Characteristic impedance, $Z_0 = 300$ ohms

∴ The normalised load impedance is given by:

$$z_L = \frac{Z_L}{Z_0}$$

$$= \frac{240 - j180}{300}$$

$$= 0.8 - j0.6$$

On the Smith chart in Fig. 2.12, locate the normalised load impedance at point A. With the centre at point O and radius OA draw the circle of voltage standing wave ratio.

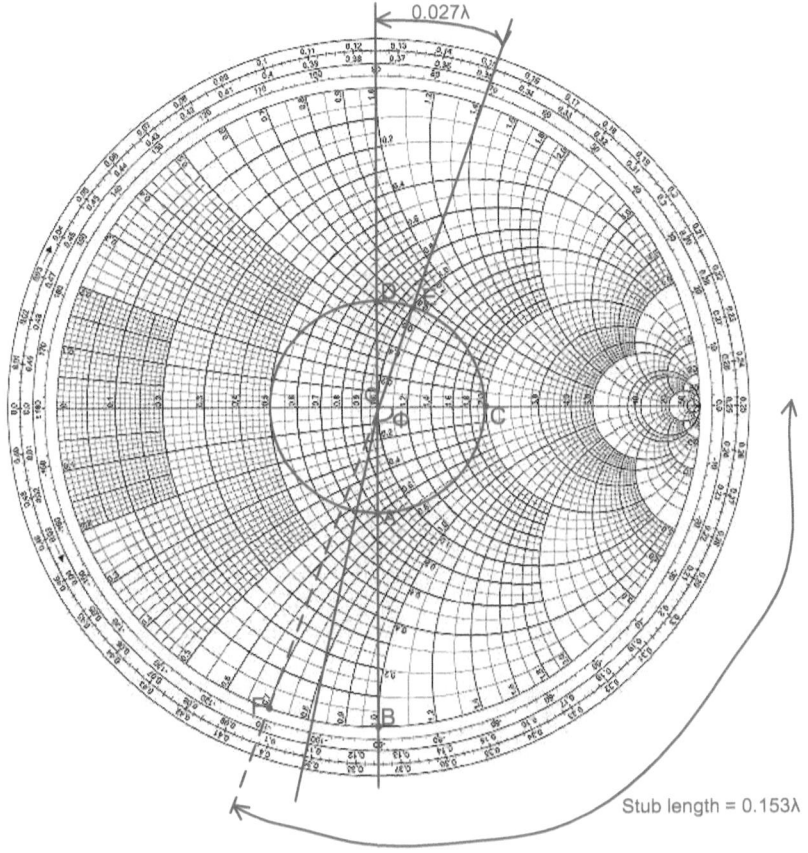

Fig. 2.12: Smith chart solution of Example 7

a) The voltage standing wave ratio is read off the chart at point C as 2.0.

b) The reflection coefficient is represented by $\dfrac{|OA|}{|OB|} \angle \phi$.

Reflection coefficient, $k = \dfrac{2.6}{7.9} \angle -90°$

$= 0.33 \angle -90°$

c) The normalized load admittance is at point D = $0.82 + j0.6$.

The nearest point of D to $y = 1 + jb$ is point E at $1 + j0.7$.

The distance of this point from the load, D to E, is $(0.152 - 0.125)\lambda = 0.027\lambda$.

Hence, the stub will be placed 0.027λ from the load and will have to tune out $b = 0.7$.

The normalised susceptance of the stub is -0.7.

Characteristic susceptance $Y_0 = \dfrac{1}{300} = 0.0033\,\text{S}$.

Hence, the susceptance of the stub will be

$Y = -0.7 \times 0.0033 = -0.0023\,\text{S}$.

Starting from the short-circuit admittance point $\infty - j\infty$ on the Smith chart and travelling clockwise around the rim of the chart, the point $0 - j0.7$ is reached. This is point F in Fig. 2.12.

The distance of F from the short-circuit admittance point is given by:

Stub Length $= (0.403 - 0.25)\lambda = 0.153\lambda$.

2.11 Questions on Transmission Lines

Question 1

The input impedance of a cable, 10km long, is $2600 + j1280$ ohms with the far end open-circuited and $220 - j140$ ohms with the far end short-circuited. The frequency of operation is 800 Hz. Calculate the primary constants of the line.

Answer: R=37.2 ohms/km; $G = 10.9 \times 10^{-6}$ S/km; L=51.3mH/km; C=0.069µF/km

Question 2

A television receiving line of negligible loss is one third of a wavelength long and has characteristic impedance of 100 ohms. The detuned receiver acts as a load of $100 + j100$ ohms. Calculate:

   a) The voltage reflection coefficient
   b) The voltage standing wave ratio,
   c) The voltage transmission coefficient,
   d) The input impedance,

Answer: $0.447 \angle 63.43°$; 2.62; $1.265 \angle 18.43°$; $38.23 - j2.56$

Question 3

An unknown load impedance connected to a 6GHz, 50-ohms transmission line causes a voltage standing wave ratio of 3.
If the first voltage minimum occurs 1.5 cm from the load plane, determine:

a) The magnitude and phase of the voltage reflection coefficient

b) The value of the unknown load impedance.

Answer: $0.5\angle 36°$; $85.15 + j66.76$ ohms

Question 4

A telephone line, 10km long, has the following constants per loop km:

R=196 ohms, L=7.1mH, C=0.069µF, G is negligible.

The line is short-circuited at the far end and a voltage of 10 volts applied at the sending end. Calculate the value of the current at the far end and its phase in relation to the applied voltage.

Answer: $4.06\angle -113.1°$

Question 5

An unknown load impedance connected to a 50MHz, 50-ohms transmission line causes the input impedance of the line to be $25 - j35$ ohms. Given that the length of the line is 3.5 m and the velocity of propagation is 3×10^8 m/s determine, using a Smith Chart:

a) The value of the load impedance

b) The magnitude and phase of the voltage reflection coefficient

c) The voltage standing wave ratio.

d) The position of the voltage minimum

Answer: $76.0 - j70.3$ ohms; $0.52\angle 40.6°$; 3.2; 1.16m from the load.

Question 6

A load impedance of $450 - j600$ ohms at 10 MHz is connected to a 300 ohms line. Obtain, using a Smith Chart: The position and length of a short-circuited stub designed to match this load to the line.

Answer: 3.9m; 2.55m

3
Fiber-Optics

3.1 Introduction

For thousands of years man has used light to communicate. Even in these days of satellite communication ships still carry powerful lamps for signaling at sea. Light is a form of electromagnetic energy. It travels at its fastest in a vacuum at 3×10^8 m/s. This speed decreases as the medium through which the light travels becomes denser.

In 1870 an Irish Physicist, John Tyndall gave a public demonstration of an experiment which shows that light could be guided by a water jet. This experiment gave birth to a revolution in communication technology [3].

3.2 Applications of Fiber-Optics

The fact that light can be guided has led to a number of interesting applications such as road signs, endoscopes, lighting hazardous areas, lighting ships at sea, and telecommunications.

3.2.1 Road Signs

A single light source can be used to power many optic fibers. This technique is used in traffic signs to indicate speed limits, lane closures, etc.

3.2.2 Endoscopes

Doctors use a bundle of very thin fibers, each carrying a single light level, to examine the insides of their patients. These are called endoscopes.

3.2.3 Lighting Hazardous Areas

A safe way to illuminate a tank containing an explosive gas is to use a light source situated a safe distance away from the tank and to transmit the light along an optic fiber. The light emitted from the end of the fiber would not have sufficient power to ignite the gas.

3.2.4 Lighting Ships at Sea

The marine environment is very hostile to electrical installations. The salt water is highly corrosive to the brass and copper used to make the connections to the lights. In this environment, optical fibers have much to offer in terms of reliability, visibility and ease of maintenance. A single light source can be used to illuminate many different locations at the same time.

3.2.5 Telecommunications

A fiber optic system using a glass fiber is capable of carrying light over long distances. By converting an input signal into short flashes of light, the optical fiber is able to carry complex information over distances of more than a hundred kilometers without additional amplification. This is at least five times better than the distances attainable using the best copper coaxial cable.

3.3 Light Wave Spectrum

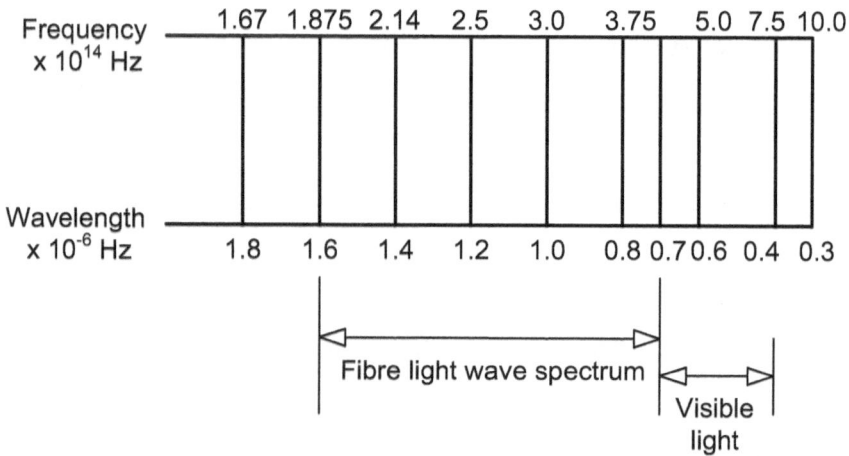

Fig. 3.1: Light wave spectrum

Fig. 3.1 shows the light wave spectrum which is just a part of the Electromagnetic spectrum. Fiber optics covers from about 0.7×10^{-6} m to about 1.6×10^{-6} m, while visible light covers from about 0.4×10^{-6} m to about 0.7×10^{-6} m [1].

3.4 Refraction

Light travels in a straight line as long as it is moving through a single uniform substance. If a ray of light traveling through one substance suddenly enters another (more or less dense) substance its speed changes abruptly causing the ray to change direction. This is why a straw sticking out of a glass of water appears bent or even broken.

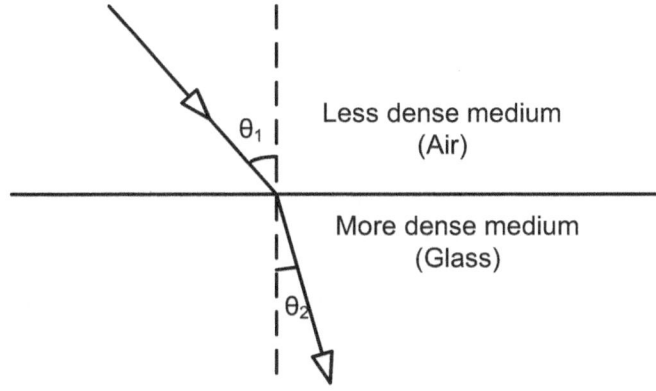

Fig. 3.2: Light from less dense medium to denser medium

Fig. 3.2 illustrates the refraction that occurs when light travels from a less dense medium to a denser medium. Here, the angle of refraction θ_2 is less than the angle of incidence θ_1.

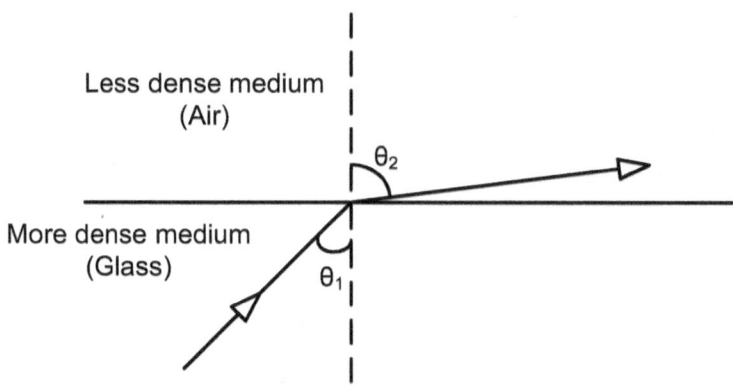

Fig. 3.3: Light from denser medium to less dense medium

Fig. 3.3 illustrates the refraction that occurs when light travels from a denser medium to a less dense medium. In this case, the angle of refraction θ_2 is more than the angle of incidence θ_1. Fiber optic technology takes advantage of this property to control the propagation of light through the fiber channel.

3.5 Critical Angle

When light is moving from a denser medium to a less dense medium the angle of refraction increases as the angle of incidence is increased.

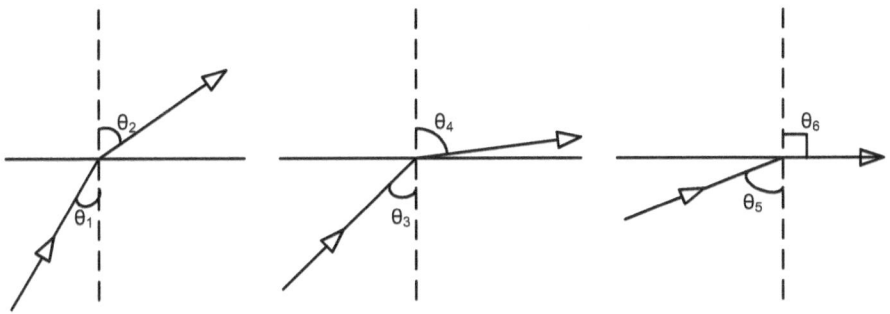

Fig. 3.4: Effect of increasing the angle of incidence

At a particular angle of incidence, θ_5 the angle of refraction becomes $90°$. At this point the refracted beam lies along the horizontal. The incident angle θ_5 at this point is known as the critical angle. This phenomenon is illustrated in Fig. 3.4.

3.6 Snell's Law

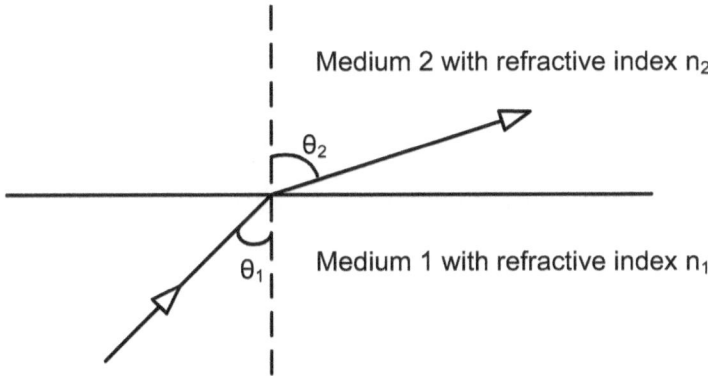

Fig. 3.5: Refraction at a junction of two media

When light passes through fiber, Snell's law gives the relationship between the incident and refracted rays as follows:

$$n_1 \sin \theta_1 = n_2 \sin \theta_2 \quad \text{............(3.1)}$$

where n_1 = refractive index in medium 1,

θ_1 = angle of incidence in medium 1,

n_2 = refractive index in medium 2, and

θ_2 = angle of refraction in medium 2.

At the critical angle of incidence θ_c, $\theta_2 = 90°$.

Hence, $n_1 \sin \theta_c = n_2 \sin 90°$.

$$n_1 \sin \theta_c = n_2$$

$$\sin \theta_c = \frac{n_2}{n_1}$$

$$\therefore \theta_c = \sin^{-1}\left(\frac{n_2}{n_1}\right) \quad \text{............(3.2)}$$

3.7 Reflection

When the angle of incidence becomes greater than the critical angle then Total Internal Reflection (TIR) takes place. This is illustrated in Fig. 3.6.

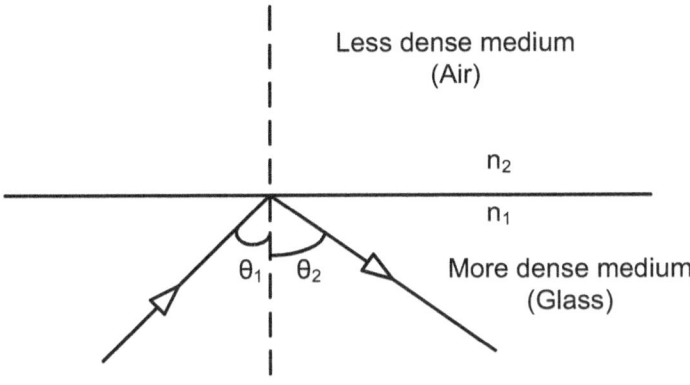

Fig. 3.6: Reflection in fiber

Here the angle of incidence, θ_1 is always equal to the angle of reflection, θ_2. Optical fiber uses reflection to guide light through a channel. From Equation (3.2) the critical angle for glass is given by:

$$\theta_c = \sin^{-1}\left(\frac{1}{1.5}\right) = 41.8°.$$

Hence for light to be propagated through this class of glass, the angle of incidence must be greater than $41.8°$.

3.8 Problem of Power Loss

When there is dirt such as water or grease on a glass, the dirt changes the refractive index of the material surrounding the glass at the point of the dirt.

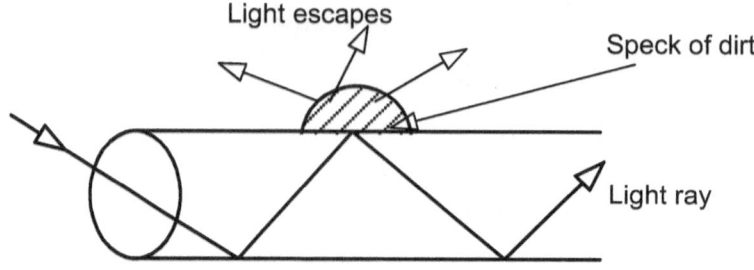

Fig. 3.7: Power loss due to contamination

Previously it was air with refractive index of 1. Now it is grease or water with a refractive index greater than 1. This will locally increase the critical angle and some of the light will find itself approaching the surface at an angle less than the new critical angle. Some of the light will be able to escape as illustrated in Fig. 3.7. This constitutes a loss.

3.9 Solution to the Problem of Power Loss caused by dirt

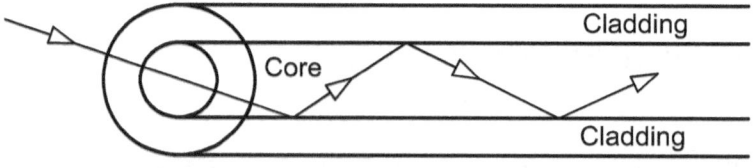

Fig. 3.8: Core surrounded with cladding

The problem of losses due to dirt is solved by covering the fiber with another layer of glass called a cladding, as shown in Fig. 3.8. The core and cladding form a single solid fiber of glass. The cladding must have a refractive index less than the core. The difference in refractive index between the core and the cladding is about 1% in most telecommunication fibers. A 1% difference corresponds to a critical angle of about 82^0.

3.10 Propagation of light along the fiber

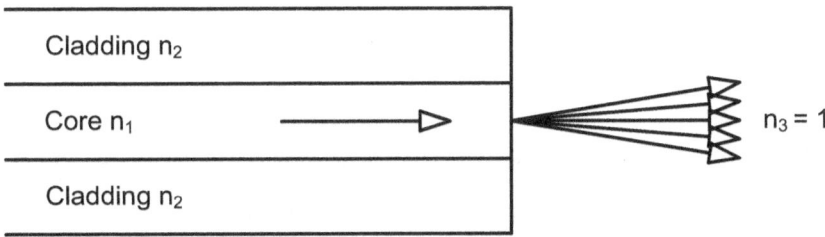

Fig. 3.9: Light spreading at the end of the fiber

When the light reaches the end of the fiber it spreads out as illustrated in Fig. 3.9. This means that if the light is shone at the same angle then the light will propagate along the fiber. For propagation to take place the angle of incidence must be greater than the critical angle.

3.11 Numerical Aperture

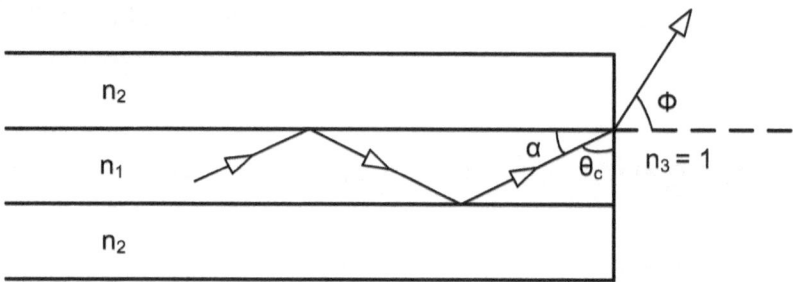

Fig. 3.10: Total internal reflection

The ray shown in Fig. 3.10 represents the ray with angle of incidence equal to the critical angle θ_c. If the core had continued then, from Snell's law we have:

$n_1 \sin \theta_c = n_2 \sin 90°$.

$$\therefore \sin \theta_c = \frac{n_2}{n_1} \quad \ldots\ldots\ldots\ldots\ldots\ldots\ldots\ldots\ldots\ldots\ldots\ldots\ldots\ldots\ldots\ldots\ldots\ldots\ldots\ldots\ldots\ldots (3.3)$$

Considering the refraction of light at the end of the core, the following relation based on Snell's law also holds:

$n_1 \sin \alpha = n_3 \sin \phi$

But $n_3 = 1$,

$\therefore \sin \phi = n_1 \sin \alpha$.

But $\alpha = 90° - \theta_c$,

$\therefore \sin \phi = n_1 \sin(90° - \theta_c)$

$$= n_1(\sin 90° \cos\theta_c - \cos 90° \sin\theta_c)$$

$$= n_1 \cos\theta_c$$

$$= n_1\sqrt{(1-\sin^2\theta_c)} \quad \dotfill (3.4)$$

$$\therefore n_1 = \frac{\sin\phi}{\sqrt{(1-\sin^2\theta_c)}} \quad \dotfill (3.5)$$

From Equation (3.4), $\sin\phi = n_1\sqrt{(1-\sin^2\theta_c)}$.

But from Equation (3.3), $\sin\theta_c = \dfrac{n_2}{n_1}$.

$$\therefore \sin\phi = n_1\sqrt{1-\left(\frac{n_2}{n_1}\right)^2}$$

$$= \sqrt{(n_1^2 - n_2^2)}.$$

$\sin\phi$ is defined as the Numerical Aperture (NA) of the fiber.

Hence, Numerical Aperture (NA) is given by:

$$NA = \sqrt{(n_1^2 - n_2^2)} \quad \dotfill (3.6)$$

Numerical Aperture is also defined as the sine of the half angle over which the fiber can accept light rays.

From Equation (3.5), we have:

$$n_1 = \frac{\sin\phi}{\sqrt{(1-\sin^2\theta_c)}}.$$

But from Equation (3.3), $n_2 = n_1 \sin\theta_c$.

$$\therefore n_2 = \frac{\sin\phi \sin\theta_c}{\sqrt{(1-\sin^2\theta_c)}} \quad\quad\quad\quad\quad\quad\quad\quad\quad\quad\quad\quad\quad\quad\quad\quad (3.7)$$

3.12 Cone of Acceptance

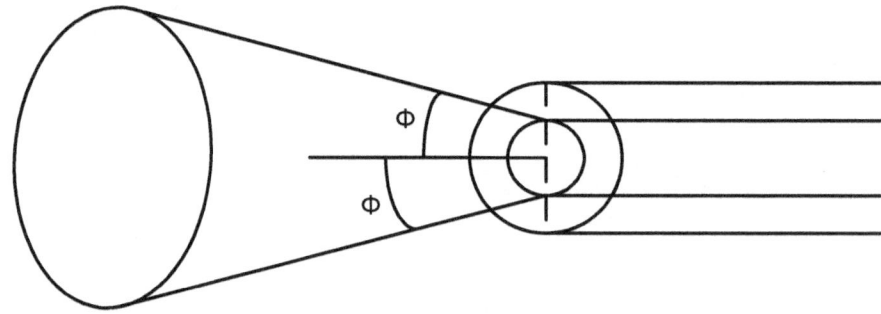

Fig. 3.11: Cone of acceptance

The angle ϕ shown in Fig. 3.11 represents the maximum angle within which the light is accepted into the core and is able to travel along the fiber. The total angle of acceptance is 2ϕ.

3.13 Modes in Fiber Optics

Mode is a description of the propagation of energy through a medium. The number of modes supported by a single fiber can be as low as one or as high as 100,000. A fiber can provide a path for one light ray or for 100,000 light rays [1],

The number of modes is given by:

$$N = \frac{1}{2}\left(D_c \times NA \times \frac{\pi}{\lambda}\right)^2 \quad\quad\quad\quad\quad\quad\quad\quad\quad\quad\quad\quad\quad\quad\quad (3.8)$$

where D_c = diameter of the core,

NA = numerical aperture of the fiber, and

λ = wavelength of the light source [1].

If $D_c = 50\mu m$, $n_1 = 1.48$, $n_2 = 1.46$, and $\lambda = 850$ nm,

then $NA = \sqrt{(1.48^2 - 1.46^2)} = 0.2425$.

$$N = \frac{1}{2}\left(50 \times 10^{-6} \times 0.2425 \times \frac{\pi}{80 \times 10^{-9}}\right)^2$$

$\quad = 1004.7$

$= 1004$.

Note that the number of modes computed must always be rounded down. Each of the 1004 modes could be represented by a ray being propagated at its own characteristic angle. Since the angles are different the different modes will take different times to cover a given distance. This leads to modal dispersion.

3.14 Multimode Step Index Fiber

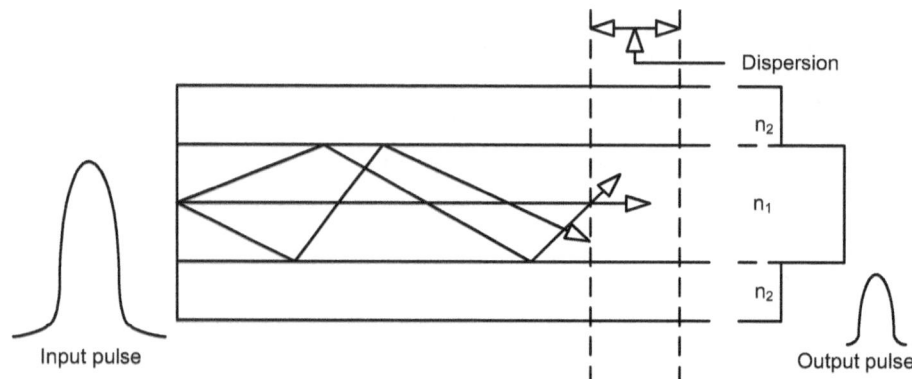

Fig. 3.12: Multimode step index fiber

The multimode step-index fiber has a core diameter of from 100 to 970μm. With this large core diameter there are many paths through which light can travel (multi-mode). The light ray traveling the straight path through the centre reaches the end before the other rays which follow a zig-zag path. The difference in the length of time it takes the various light rays to exit the fiber is called modal dispersion. This is a form of signal distortion which limits the bandwidth of the fiber.

3.15 Problem of Modal Dispersion

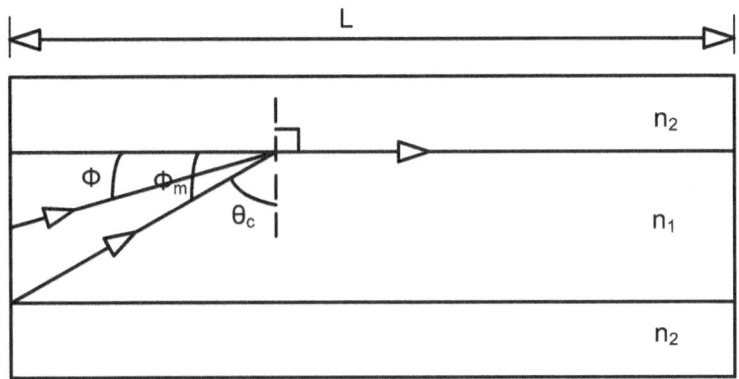

Fig. 3.13: Modal dispersion in fiber optics

For total internal reflection,

$n_1 \sin \theta_c = n_2$.

Alternatively, $\sin \theta_c = \dfrac{n_2}{n_1}$.

But $\sin \theta_c = \cos \phi_m$

$\therefore \cos \phi_m = \dfrac{n_2}{n_1}$

ϕ_m represents the maximum angle the light ray makes with the fiber axis for total internal reflection to take place.

Consider a short pulse of light launched into a fiber of length L at an angle of ϕ to the axis. When $\phi = 0$, the light travels directly along the fiber core axis and emerges at the exit after a delay given by:

$$\tau_{min} = \frac{L}{v_1}$$

$$= \frac{n_1 L}{c},$$

where $v_1 = \frac{c}{n_1}$ is the velocity of light in the core, and

c is the velocity of light in air.

On the other hand light inclined at an angle ϕ to the axis follows a zig-zag path of length L_z given by:

$$L_z = \frac{L}{\cos\phi}.$$

This light will arrive at the exit after a delay given by:

$$\tau_z = \frac{L_z}{v_1}$$

$$= \frac{n_1 L}{c \cos\phi}.$$

For the extreme ray traveling at an angle ϕ_m, we have:

$$\tau_{max} = \frac{n_1 L}{c \cos\phi_m}.$$

Recall that $\cos\phi_m = \frac{n_2}{n_1}$.

$$\therefore \tau_{max} = \frac{n_1^2 L}{c n_2}.$$

Hence if a pulse of light is launched such that components enter the fiber at all angles $0 < \phi < \phi_m$, the pulse arrives at the exit spread out in time by an amount given by:

$$\Delta \tau = \tau_{max} - \tau_{min}$$

$$= \frac{n_1^2 L}{c n_2} - \frac{n_1 L}{c}$$

$$= \frac{n_1 L}{n_2 c}(n_1 - n_2).$$

This pulse spreading places a limit on the rate at which pulses may be transmitted over a given fiber of length L. The maximum pulse rate is given by:

$$f_{max} = \frac{1}{\Delta \tau} = \frac{n_2 c}{n_1 L (n_1 - n_2)} \text{ Hz} \quad \text{................(3.9)}$$

Pulse spreading of this sort is characteristic of multimode step index fibers. If f_{max} is given then it places a limit on the length L, given by:

$$L_{max} = \frac{n_2 c}{n_1 f_{max} (n_1 - n_2)} \text{ m.} \quad \text{................(3.10)}$$

3.16 Solution to Modal Dispersion

The problem of modal dispersion (pulse spreading) can be solved in either of two ways, viz.: using multimode graded index fiber, or using single mode step index fiber.

3.16.1 Multimode Graded Index Fiber

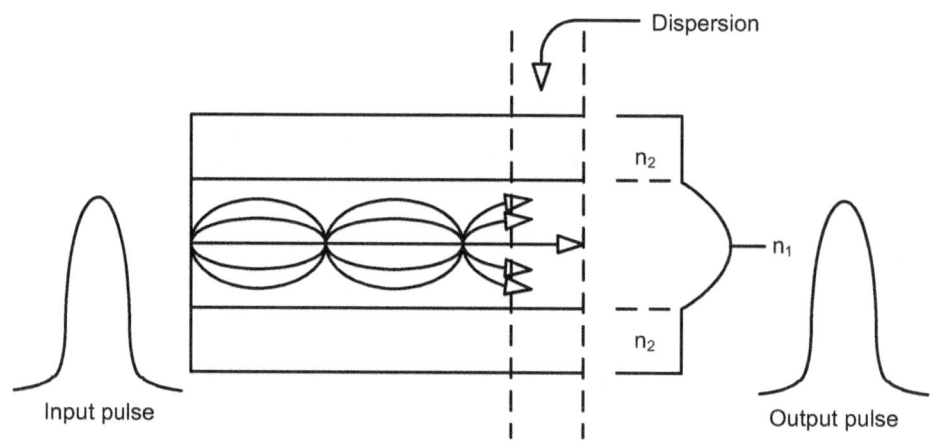

Fig. 3.14: Multimode graded index fiber

Here the core is made with the maximum refractive index at the centre. The index then reduces gradually until it reaches the inner edge of the cladding. The net effect is that the ray which travels the shortest distance will have the lowest speed. The rays which take the zig-zag path travel faster. The net effect is that the light rays arrive at the exit point at almost the same time, thus reducing modal dispersion. A typical graded index fiber has core diameter ranging from 50 to 85 µm and a cladding diameter of 125 µm.

3.16.2 Single Mode Step Index Fiber

Single mode step index fibers are the most widely used in today's wide-band communication systems. The core diameter is reduced to such a level that it can support only one mode.

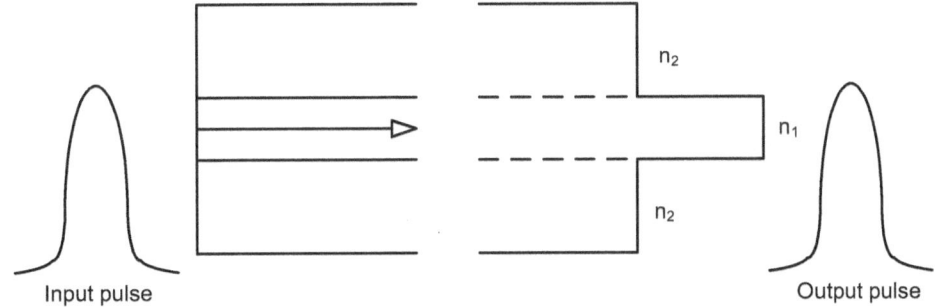

Fig. 3.15: Single mode step index fiber

Since light ray can travel on only one path, modal dispersion is zero. The core diameter ranges from 5 to 10 µm. The standard cladding diameter is 125 µm. The extra cladding thickness tends to set an overall fiber size standard and makes the fiber less fragile.

If $D_c = 5$µm, $n_1 = 1.48$, $n_2 = 1.47$, and $\lambda = 1.5$ µm,

then $NA = \sqrt{(1.48^2 - 1.47^2)} = 0.1718$.

$$N = \frac{1}{2}\left(D_c \times NA \times \frac{\pi}{\lambda}\right)^2$$

$$= \frac{1}{2}\left(5\times10^{-6} \times 0.1718 \times \frac{\pi}{1.5\times10^{-6}}\right)^2$$

$$= 1.62$$

$$= 1$$

3.17 Cut-off Wavelength

The condition for single mode transmission is given by:

$$D_c < \frac{2.4\lambda}{\pi\sqrt{n_1^2 - n_2^2}} \quad \ldots (3.11)$$

If the core diameter is larger, then the fiber can carry two or more modes. From Equation (3.11), we have:

$$\lambda > \frac{\pi D_c \sqrt{n_1^2 - n_2^2}}{2.4} \quad \ldots (3.12)$$

Equation (3.12) shows that a specific core diameter transmits light in a single mode only at a wavelength longer than a value called the cut-off wavelength given by:

$$\lambda_c = \frac{\pi D_c \sqrt{n_1^2 - n_2^2}}{2.4} \quad \ldots (3.13)$$

At wavelengths shorter than λ_c, additional modes will appear.

3.18 Cable Composition

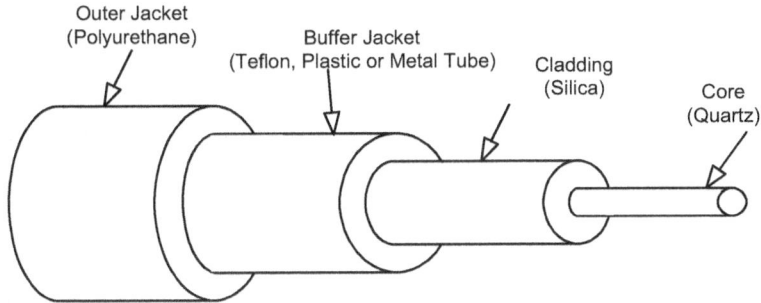

Fig. 3.16: Fiber construction

A core, made of quartz is surrounded by a cladding made of silica. This constitutes the fiber. The fiber is often covered with a buffer to protect it from moisture. Finally the entire cable is encased in an outer jacket. The choice of material for the buffer and outer jacket depends on where the cable is to be installed.

3.19 Fiber Sizes

Optical fibers are defined by the ratio of the diameter of their core to the diameter of their cladding, both expressed in microns.

Fiber Type	Core (microns)	Cladding (microns)
8.3/125	8.3	125
50/125	50	125
62.5/125	62.5	125
100/140	100	140

TABLE 3.1: Fiber Types

3.20 Light Sources and Detectors for Optical Cables

The purpose of fiber optic cable is to contain and direct a beam of light from the source to the target. For transmission to occur the sending device must be equipped with a light source and the receiving device with a photo sensitive cell – photo diode capable of translating the received light into current usable by a computer. Light sources can be either an Injection Laser Diode (ILD) or a Light Emitting Diode (LED). Received light can be converted into an electrical signal by a PIN Diode or an Avalanche Photo Diode (APD).

3.20.1 Injection Laser Diode

A laser provides a light of fixed wavelength which can be in the visible region around 635 nm or in any of the three infrared windows namely: 850 nm, 1,300 nm or 1,550 nm. The 850 nm is used for medium distance data transmission. The 1,300 nm and 1,550 nm windows are used for long distance telecommunications. Lasers have the following advantages:

- They can be focused to a very narrow range allowing control over the angle of incidence.
- They can be modulated to the rate of gigabits per second.
- They preserve the character of the signal over long distances.

The main disadvantage of lasers is that they are expensive.

3.20.2 Light Emitting Diodes (LEDs)

LEDs can provide light output in the visible spectrum as well as in the 850 nm, 1,300 nm, and 1,550 nm windows. The advantages of LEDs are as follows:

- They are cheaper than ILDs.
- They are more reliable than ILDs.
- They are not temperature sensitive and can operate over a large temperature range without the use of temperature control circuits.

The disadvantages of LEDs are as follows:

- They have slower switching speed (only a few hundred megabits per second) compared to ILDs.
- They produce unfocussed light, hence they are limited to short distance use.

3.20.3 PIN Diodes

A PIN diode is the most popular method of converting the received light into an electronic signal. It is usual for a PIN diode to have an amplifier built into the module to provide a higher output signal level.

3.20.4 Avalanche Photo Diode

Avalanche photo diodes have the advantage of a good output at low light levels. They have wide dynamic range. They have the following disadvantages:
- High noise level.
- They cost more than PIN diodes.
- They require higher operating voltages.
- The gain decreases with increase in temperature.

3.21 Advantages of Optical Fiber

Optical fiber has the following advantages:
- They are immune to electrical interference. External light is blocked by the outer jacket.
- No cross-talk.
- They have low losses, typically 0.2 dB/km. This means they can be used over long distances before repeaters become necessary.
- They have improved bandwidth in the Gigahertz frequency range. Copper cables are restricted to 500 MHz.
- They are lightweight and non-corrosive. Hence they are useful for aircraft and automotive applications.

- They are insulated hence they are safe for use in high voltage areas.
- They offer security of data since they do not radiate electromagnetic signals.
- One fiber can send a signal whereas copper requires two wires, one of which is needed as a return path to complete the electrical circuit.

3.22 Disadvantages of Optical Fiber

Optical fiber has the following disadvantages:
- The manufacturing cost is high. A laser light source can be ten times more costly than signal generators.
- The cost of installation and maintenance is high.
- They are fragile.

3.23 Connecting Optical Fibers

There are three things to consider when connecting optic fibers:
- The fibers must be compatible types.
- The ends of the fibers must be brought together in close proximity.
- The fibers must be accurately aligned.

3.23.1 Compatibility

If the fibers are not compatible then the degree of loss will depend on the direction of travel of the light along the fiber.

Losses Due To Unequal Core Sizes

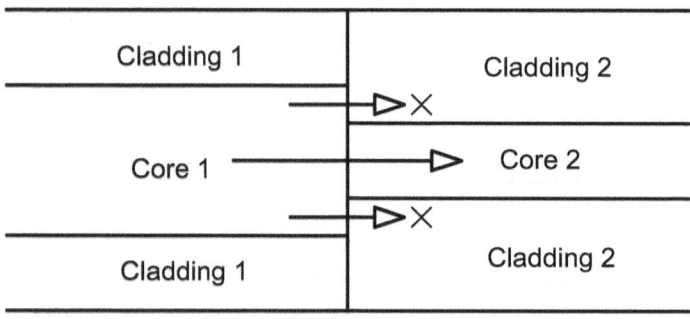

Fig. 3.17: Large core to small core

If two fibers with different core sizes are connected together as shown in Fig. 3.17 and light is travelling from the large core to the small core, there will be loss because some light from the large core will not be able to enter the smaller core.

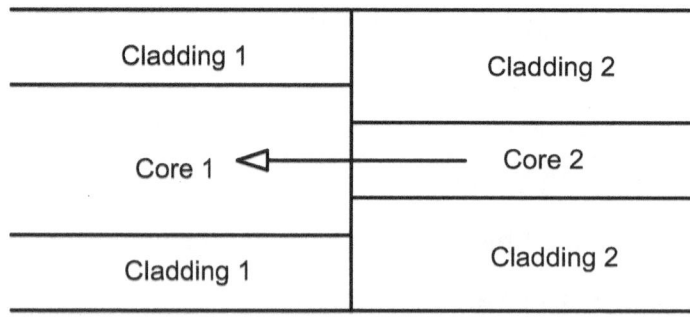

Fig. 3.18: Small core to large core

However if light is travelling from the small core to the large core as depicted in Fig. 3.18, there will be no loss.

Receive Fiber Core Size (μm)	Launch Fiber Core Size (μm)		
	9	50	62.5
9	0	14.9dB	16.8dB
50	0	0	1.9dB
62.5	0	0	0

TABLE 3.2: Losses due to unequal core sizes. [3]

TABLE 3.2 shows the losses due to unequal core sizes. The loss is given by:

$$\text{Loss} = -10\log_{10}\left(\frac{D_{c_{receive}}}{D_{c_{launch}}}\right)^2 \text{ dB}, \quad\quad\quad\quad\quad\quad\quad\quad\quad\quad (3.14)$$

where $D_{c_{receive}}$ = core diameter of receive fiber cable, and

$D_{c_{launch}}$ = core diameter of launch fiber cable [3].

This applies only when the diameter of the launch fiber is greater than that of the receiving fiber, otherwise there are no losses.

Losses due to unequal Numerical Aperture

If $NA_{receive} \geq NA_{launch}$, then there will be no losses.

Recall that the numerical aperture determines the cone of acceptance. Consider a launch fiber with $NA = 0.25$, cone angle $= 14.5°$ and a receive fiber with $NA = 0.2$, cone angle $= 11.5°$, as illustrated in Fig. 3.19.

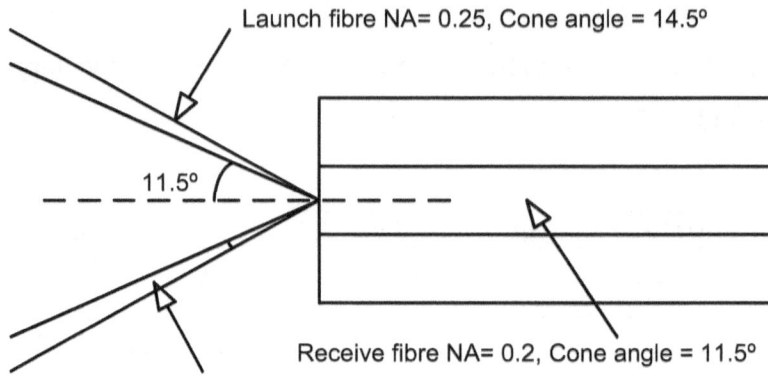

Fig. 3.19: Losses due to unequal numerical aperture

Some of the light from the launch fiber cannot enter the receive fiber.

Receive Fiber NA	Launch Fiber NA		
	0.1	0.2	0.275
0.1	0	6dB	8.79dB
0.2	0	0	2.8dB
0.275	0	0	0

TABLE 3.3: Losses due to unequal numerical aperture. [3]

TABLE 3.3 shows the losses due to unequal numerical aperture. The loss is given by:

$$\text{Loss} = -10\log_{10}\left(\frac{NA_{receive}}{NA_{launch}}\right)^2 \text{ dB}, \quad\quad\quad\quad\quad (3.15)$$

where $NA_{receive}$ = numerical aperture of receive fiber cable, and

NA_{launch} = numerical aperture of launch fiber cable [3].

This applies only when the numerical aperture of the launch fiber is greater than that of the receiving fiber, otherwise there are no losses. If the core sizes and the *NA* are different then the losses add.

3.23.2 Gap Loss

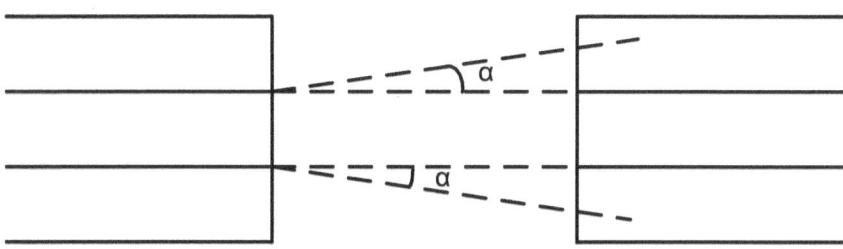

Fig. 3.20: Gap loss

Gap loss refers to the loss that occurs when the ends of two connected fibers are not in close proximity, as illustrated in Fig. 3.20. The degree of loss is of the order of 0.5dB when the ends of the fiber are separated by a distance equal to a core diameter. The loss is reduced by the use of index matching gel which is added in the joint to make the fiber core appear continuous. The gel should have a refractive index similar to that of the core.

3.23.3 Alignment Problems

Fig. 3.21 shows the loss that occurs when two connected fibers are misaligned laterally. This is similar to the loss due to differences in the core size.

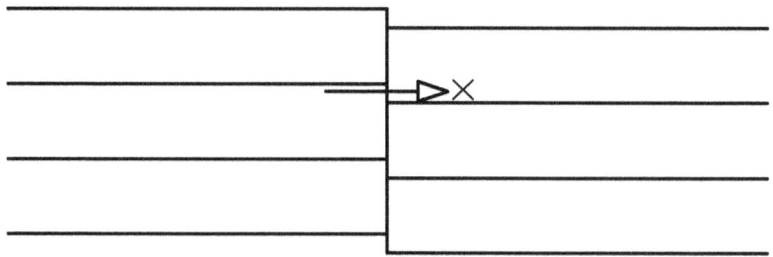

Fig. 3.21: Loss due to lateral misalignment

A misalignment of one quarter the core diameter will cause a loss of 1.5dB. Thereafter the loss increases rapidly in a nonlinear manner.

3.24 Splicing of Optical Fibers

A splice is a device or process used to connect fibers. The basic requirements of splices are low loss and accurate alignment. A splice can be used to extend cable length or to repair a break. There are two basic types of splices namely: Fusion and Mechanical.

3.24.1 Fusion Splicing

The fibers are aligned and joined by electric arc welding. They are more permanent, and have lower loss, typically less than 0.05dB. However, they are more expensive and are not suitable for field service repair.

3.24.2 Mechanical Splicing

The fibers are stripped, cleaned and cleaved. It does not require power supply. It is reusable. It can be done in a few minutes. However, it has high loss of between 0.1 and 0.3 dB per connection.

3.25 Examples on Fiber Optics

Example 1

A multi-mode step index fiber has a core refractive index of 1.52, a cladding refractive index of 1.51 and a loss of 5dB/km.
  a) From the point of view of pulse spreading what is the maximum fiber length allowable for data transmission at 2 Mbits /s?
  b) What would be the loss of a fiber of this length?
  c) If 0.5mW of optical power is launched into the fiber, determine the output power level,

Solution

a) Modal dispersion is given by:

$$\Delta\tau = \frac{n_1 L}{n_2 c}(n_1 - n_2).$$

The maximum frequency is given by:

$$f_{max} = \frac{1}{\Delta\tau} = \frac{n_2 c}{n_1 L(n_1 - n_2)}$$

$$L_{max} = \frac{n_2 c}{n_1 f_{max}(n_1 - n_2)}$$

$$= \frac{1.51 \times 3 \times 10^8}{1.52 \times 2 \times 10^6 (1.52 - 1.51)}$$

$$= \frac{1.51 \times 300}{1.52 \times 2 \times 0.01}$$

$$= 14901\,\text{m}$$

$$= 14.9\,\text{km}.$$

b) Loss = 5dB/km.

Total loss = $5 \times 14.9 = 74.5\,\text{dB}$

c) Output power = P_o

Input power = $P_i = 0.5\,\text{mW}$

$$\text{Loss(dB)} = 10\log\left(\frac{P_o}{P_i}\right)$$

$$\therefore -74.5 = 10\log\left(\frac{P_o}{P_i}\right)$$

$$\log\left(\frac{P_o}{P_i}\right) = -7.45$$

$$\left(\frac{P_o}{P_i}\right) = 10^{-7.45} = 3.55 \times 10^{-8}$$

$$P_o = 3.55 \times 10^{-8} \times 0.5$$

$$P_o = 1.78 \times 10^{-8}\,\text{mW} = 1.78 \times 10^{-11}\,\text{W}.$$

Example 2

A fiber optic cable has a core diameter of 50×10^{-6} m and refractive index of 1.5. The cladding is of refractive index 1.48. Calculate:

a) The critical angle.
b) The acceptance angle.
c) The number of modes that would occur if a light source of wavelength 850×10^{-9} m is used.

Solution

Core refractive index, $n_1 = 1.5$

Cladding refractive index, $n_2 = 1.48$

Core diameter, $D_c = 50 \times 10^{-6}$ m

Wavelength, $\lambda = 850 \times 10^{-9}$ m

a) From Snell's law:

$n_1 \sin \theta_c = n_2 \sin 90°$, where θ_c is the critical angle.

$$\therefore \sin \theta_c = \frac{n_2}{n_1} = \frac{1.48}{1.5} = 0.987.$$

$\therefore \theta_c = 80.6°.$

b) The numerical aperture is given by:

$$NA = \sqrt{(n_1^2 - n_2^2)} = \sqrt{1.5^2 - 1.48^2} = 0.24413$$

Acceptance angle $= \sin^{-1} NA = 14.12°$

c) The number of modes is given by:

$$N = \frac{1}{2}\left(D_c \times NA \times \frac{\pi}{\lambda}\right)^2$$

$$= \frac{1}{2}\left(50 \times 10^{-6} \times 0.2441 \times \frac{\pi}{850 \times 10^{-9}}\right)^2$$

$= 1017.69$

$= 1017$ modes

Example 3

A multi-mode step-index fiber has a core in which the critical angle is $75.2°$, the acceptance angle is $22.6°$ and the core diameter is 62.5×10^{-6} m. Calculate:

a) The refractive indices of the core and cladding.

Fiber-Optics

b) The loss that would occur if this cable is spliced to another cable of core diameter 50×10^{-6} m.

Solution

Critical angle, $\theta_c = 75.2°$.

Acceptance angle, $\phi = 22.6°$.

Diameter of core 1, $D_{c_1} = 62.5 \times 10^{-6}$ m

Diameter of core 2, $D_{c_2} = 50 \times 10^{-6}$ m

a) Let the refractive index of the core be n_1.

Let the refractive index of the cladding be n_2

From Snell's law:

$$\sin \theta_c = \frac{n_2}{n_1}$$

$$\therefore n_2 = n_1 \sin \theta_c$$

Also, $\sin \phi = \sqrt{(n_1^2 - n_2^2)}$.

$$\therefore \sin^2 \phi = n_1^2 - n_2^2$$

But $n_2 = n_1 \sin \theta_c$.

$$\therefore \sin^2 \phi = n_1^2 - n_1^2 \sin^2 \theta_c$$

$$\sin^2 \phi = n_1^2 (1 - \sin^2 \theta_c)$$

$$\therefore n_1 = \sqrt{\frac{\sin^2 \phi}{1 - \sin^2 \theta_c}}$$

$$= \frac{\sin\phi}{\sqrt{1-\sin^2\theta_c}}$$

$$= \frac{\sin 22.6°}{\sqrt{1-\sin^2 75.2°}}$$

$$= \frac{0.3843}{0.2554}$$

$$\therefore n_1 = 1.5.$$

$$n_2 = n_1 \sin\theta_c = 1.5\sin 75.2° = 1.45.$$

b) The loss due to unequal core sizes is given by:

$$\text{Loss} = -10\log_{10}\left(\frac{D_{c_{receive}}}{D_{c_{launch}}}\right)^2 = -10\log_{10}\left(\frac{D_{c_2}}{D_{c_1}}\right)^2 \text{ dB}$$

$$= -10\log_{10}\left(\frac{50}{62.5}\right)^2$$

$$= 1.94\text{dB}.$$

Example 4

Given the following data for a fiber optic cable:

Speed of light in a vacuum = 3×10^8 m/s,

Speed of light in the core of the fiber optic cable = 2×10^8 m/s, and

Speed of light in the cladding of the fiber optic cable = 2.027×10^8 m/s, calculate:

  a) The critical angle of the fiber optic cable.
  b) The numerical aperture.
  c) The modal dispersion for a meter of the cable.

d) The loss that would occur if this cable is spliced to another cable of numerical aperture 0.18.

e) How is modal dispersion eliminated?

Solution

The refractive index, n, of a material is defined as:

$$n = \frac{c}{v},$$

where c is the speed of light in a vacuum and v is the speed of light in the material.

Hence for the core, we have:

$$n_1 = \frac{c}{v_1} = \frac{3 \times 10^8}{2 \times 10^8} = 1.5.$$

For the cladding, we have:

$$n_2 = \frac{c}{v_2} = \frac{3 \times 10^8}{2.027 \times 10^8} = 1.48.$$

a) The critical angle of the fiber optic cable is given by:

$$\theta_c = \sin^{-1}\left(\frac{n_2}{n_1}\right)$$

$$= \sin^{-1}\left(\frac{1.48}{1.5}\right)$$

$$\therefore \theta_c = 80.6°.$$

b) The numerical aperture is given by:

$$NA = \sqrt{(n_1^2 - n_2^2)} = \sqrt{1.5^2 - 1.48^2} = 0.24413.$$

c) Modal dispersion is given by:

$$\Delta \tau = \frac{n_1 L}{n_2 c}(n_1 - n_2).$$

$L = 1\,\text{m}$

$$\therefore \Delta \tau = \frac{1.5 \times 1}{1.48 \times 3 \times 10^8}(1.5 - 1.48)\,\text{s/m}$$

$$= 0.0676\,\text{ns/m}$$

d) Numerical aperture of the first cable $NA_1 = 0.244$.

Numerical aperture of the second cable $NA_2 = 0.18$.

The loss due to unequal numerical apertures is given by:

$$\text{Loss} = -10\log_{10}\left(\frac{NA_2}{NA_1}\right)^2\,\text{dB}$$

$$= -10\log_{10}\left(\frac{0.18}{0.244}\right)^2$$

$$= 2.64\,\text{dB}.$$

e) Modal dispersion can be eliminated by reducing the core diameter to such a level that it can support only one mode. The core diameter for such ranges from 5 to 10μm. This is called Single Mode Step Index Fiber.

Example 5

A multi-mode step-index fiber has modal dispersion of 30 ns/km. Calculate the distance over which it could transmit a signal at 1 Gbits/s.

Solution

Modal dispersion is given by:

$$\Delta \tau = \frac{n_1 L}{n_2 c}(n_1 - n_2).$$

When $L = 1000$ m, $\Delta \tau = 30 \times 10^{-9}$ s.

$$\therefore 30 \times 10^{-9} = \frac{1000 n_1 (n_1 - n_2)}{n_2 c}$$

$$\frac{n_1(n_1 - n_2)}{n_2 c} = 3 \times 10^{-11}$$

$$f = \frac{1}{\Delta \tau} = \frac{n_2 c}{n_1 L (n_1 - n_2)}$$

$$\therefore L = \frac{n_2 c}{n_1 f (n_1 - n_2)}$$

$f = 1$ Gbits/s $= 10^9$ bits/s

$$\therefore L = \frac{1}{10^9 \times 3 \times 10^{-11}} = \frac{100}{3} = 33.3 \, \text{m}.$$

3.26 Questions on Fiber Optics

Question 1

An optical fiber has a core refractive index of 1.5 and a cladding refractive index of 1.49. Calculate:
   a) The critical angle of the fiber optic cable.
   b) The numerical aperture.
   c) The modal dispersion for a 100 m of cable.
   d) The loss that would result if this cable is spliced to another cable of numerical aperture 0.1.

Answer: 83.4°; 0.173; 3.36ns; 4.8dB

Question 2

A multi-mode step-index fiber has the following parameters:

Critical angle = 80.6°

Core diameter = 50×10^{-6} m

Wavelength of light source = 850 nm

Number of modes = 1004

Calculate:
   a) The refractive indices of the core and the cladding.
   b) The numerical aperture.
   c) The maximum fiber length allowable for data transmission at 1 Mbits/s.

Answer: 1.49; 1.47; 0.2425; 14.8 km

Question 3

A multi-mode step-index fiber has a core of 50×10^{-6} m diameter and a numerical aperture of 0.2424. Calculate:

a) The number of modes that would occur using a light source of wavelength 865 nm.

b) What loss would result if this fiber is spliced to another one whose core diameter is 9×10^{-6} m.

Answer: 1937; 14.9 dB.

Question 4

A fiber optic cable has the following parameters:

Refractive index of the core = 1.5.

Refractive index differential between the core and the cladding = 1%.

Calculate:

a) The critical angle.

b) The numerical aperture.

c) The cut-off wavelength for a core diameter of 4.8×10^{-6} m.

Answer: $81.9°$; 0.2116; 33×10^{-6} m

Question 5

A fiber optic cable has the following parameters:

Critical angle = $75.2°$

Numerical aperture = 0.384

Calculate:

a) The refractive indices of the core and cladding.
b) The maximum signaling rate attainable for fiber of length 1 Km.

Answer: 1.50; 1.45; 5.8 MHz

4
Waveguides

4.1 Introduction

The name wave guide is reserved for specially constructed hollow metallic pipes. They are used at microwave frequencies.

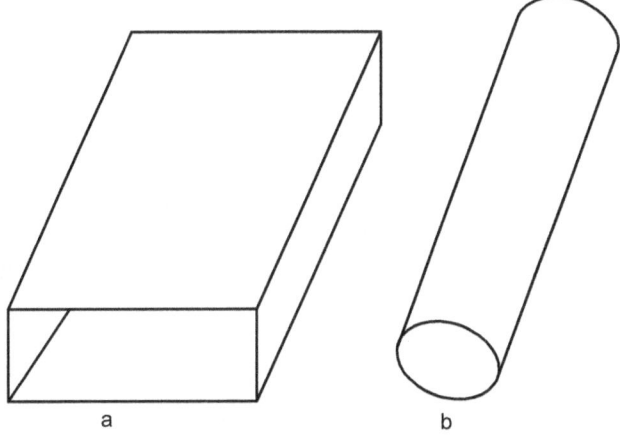

Fig. 4.1: a) Rectangular and b) Circular waveguides

Wave guides are preferred to transmission lines because they are much less lossy at the highest frequencies. It is a known fact that at very high frequencies the majority of the current flow will occur mostly along the surface of the conductor. This phenomenon, known as skin effect has led to the development of hollow conductors known as wave guides.

Waveguides with constant rectangular or circular cross-sections are normally employed, although other shapes may be used for special purposes. The walls of the guide are conductors. However, conduction of energy does not take place through the walls whose functions are to confine this energy and allow reflections to take place. Conduction takes place through the dielectric filling the wave guide. The dielectric is usually air. In discussing the behaviour and properties of wave guides it is necessary to look at electric and magnetic fields as in wave propagation instead of voltages and currents as in transmission lines.

4.2 Applications

The cross-sectional dimensions of a wave guide must be of the same order as those of a wavelength, hence use at frequencies below about 1 GHz is not normally practical. TABLE 4.1 gives some selected rectangular waveguide sizes together with their frequencies of operation. [1]

Useful Frequency Range (GHz)	Outside Dimensions (mm)	Wall Thickness (mm)	Type	Average Power Rating (kW)
1.12 – 1.70	169 x 86.6	2.0	WR 650	14600
3.95 – 5.85	50.8 x 25.4	1.6	WR 187	1700
12.40 – 18.00	17.8 x 9.8	1.0	WR 62	140
18.00 – 26.50	12.7 x 6.4	1.0	WR 42	51
60.00 – 90.00	5.1 x 3.6	1.0	WR 12	5.1

TABLE 4.1: Examples of rectangular waveguides

4.3 Advantages of Waveguides

Waveguides have the following advantages:
- They are simpler to manufacture compared to coaxial lines.
- Flashover is less likely to occur in wave guides.
- Improved power handling ability by about ten times more than for coaxial air dielectric rigid cables of similar dimensions.
- They have higher maximum operating frequencies (325 GHz compared to 18 GHz for coaxial lines).
- They have lower power losses than for comparable transmission lines.

4.4 Method of Wave Propagation in a Waveguide

Normal electromagnetic wave is Transverse – Electromagnetic, TEM. The Electric and Magnetic fields and the direction of propagation are mutually perpendicular, as depicted in Fig. 4.2.

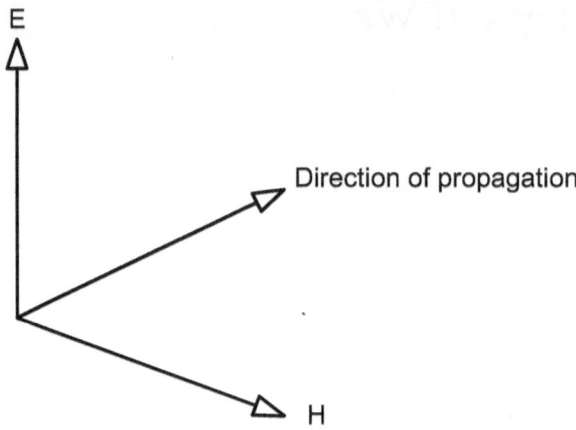

Fig. 4.2: Transverse electromagnetic wave transmission

This type of wave will not propagate in a wave guide. The electric field would be short-circuited by the walls of the wave guide. Propagation is achieved by sending the wave in a zigzag fashion as illustrated in Fig. 4.3.

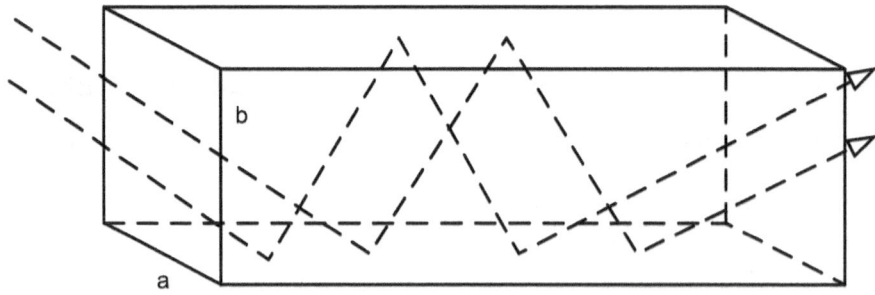

Fig. 4.3: Wave propagation in a waveguide

There are two consequences of the zigzag propagation:
- The velocity of propagation in a waveguide is less than in free space.

- The waves can no longer be TEM. Propagation by reflection requires not only a normal component but also a component in the direction of propagation for either the electric or the magnetic field.

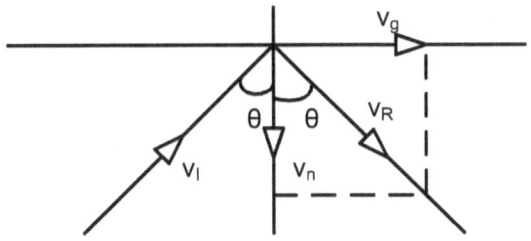

Fig. 4.4: Reflection from a conducting surface

The two different possibilities are as follows:
- Transverse Electric Mode (TE): Here the electric field lies wholly in the transverse plane perpendicular to the direction of propagation while there is a magnetic component along the direction of propagation. This type is sometimes called H-wave.
- Transverse Magnetic Mode (TM): Here the magnetic field exists only in the plane transverse to the direction of propagation and there is a component of electric field in the direction of propagation. This type is sometimes called E-wave.

4.5 Dominant Mode of Propagation

The dominant mode is the lowest possible frequency that can be propagated in a given waveguide.

$TE_{m,n}$ = Transverse Electric

$TM_{m,n}$ = Transverse Magnetic,

where m = number of half wavelengths across the waveguide width a, and

n = number of half wavelengths across the waveguide height b [1].

4.6 Plane Waves at a Conducting Surface

With reference to Fig. 4.5, waves travel from left to right.

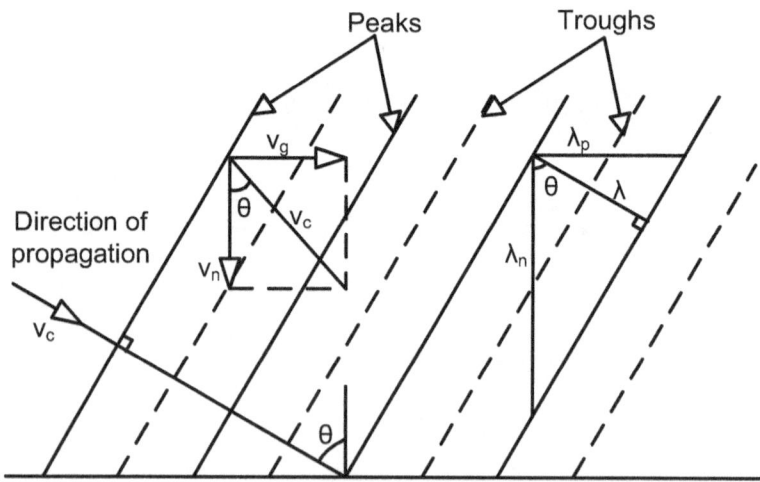

Fig. 4.5: Wavefronts incident on a perfectly conducting plane- reflections are not shown

θ = angle of incidence

v_c = actual velocity of propagation

v_g = velocity parallel to the conducting surface

v_n = velocity normal to the surface

λ = wavelength in direction of propagation (distance between two wave peaks)

λ_p = wavelength in direction parallel to the surface

λ_n = wavelength at right angle to the surface

$$v_g = v_c \sin\theta \quad \text{...(4.1)}$$

$$v_n = v_c \cos\theta \quad \text{...(4.2)}$$

Equation (4.1) shows that waves travel more slowly in waveguides than in free space.

$$\lambda_p = \frac{\lambda}{\sin\theta} \quad \text{...(4.3)}$$

$$\lambda_n = \frac{\lambda}{\cos\theta} \quad \text{...(4.4)}$$

Equations (4.3) and (4.4) show that the wavelength is greater when measured in a direction other than the direction of propagation.

4.7 Phase Velocity

Any electromagnetic wave has two velocities, the one with which it propagates and the one with which it changes phase. In free space the two velocities are the same and equal to the speed of light, v_c.

$v_c = f\lambda = 3 \times 10^8$ m/s in free space.

From Equation (4.3), we have:

$$\lambda_p = \frac{\lambda}{\sin\theta}$$

$$v_p = f\lambda_p = \frac{f\lambda}{\sin\theta}$$

Hence, $v_p = \dfrac{v_c}{\sin\theta}$ ....(4.5)

where v_p is the phase velocity.

4.8 Cut-off Wavelength

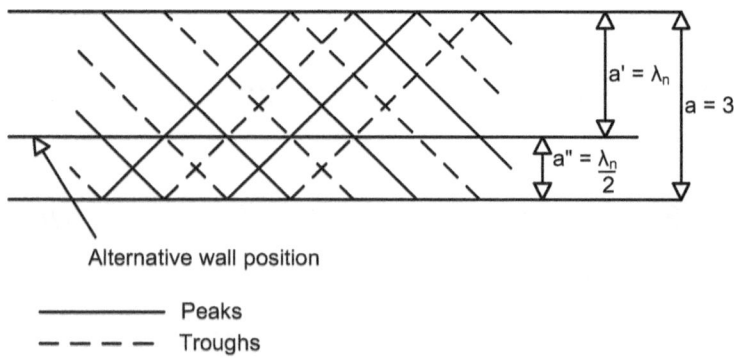

Fig. 4.6: Reflection in a parallel plane waveguide

If a second wall is added to the first one at a distance a'' from it, then it must be placed at a point where the electric intensity due to the first wall is zero. This implies that the second wall is placed at an integral number of half wavelengths away.

Hence $a = \dfrac{m\lambda_n}{2}$, ...(4.6)

where:

a = distance between the walls,

λ_n = wavelength in a direction normal to both walls, and

m = number of half wavelengths of electric intensity established between the walls.

m is an integer.

Recall that $\lambda_n = \dfrac{\lambda}{\cos\theta}$.

Hence, $a = \dfrac{m\lambda}{2\cos\theta}$.

$\therefore \cos\theta = \dfrac{m\lambda}{2a}$.

Also, $\lambda_p = \dfrac{\lambda}{\sin\theta}$

$= \dfrac{\lambda}{\sqrt{1-\cos^2\theta}}$

$\therefore \lambda_p = \dfrac{\lambda}{\sqrt{1-\left(\dfrac{m\lambda}{2a}\right)^2}}$ ....(4.7)

From Equation (4.7), it can be seen that at a particular value of λ for which $\lambda_p = \infty$ the wave can no longer propagate. This represents cut-off. At cut-off, $\lambda = \lambda_c$.

$$\therefore 1 - \left(\frac{m\lambda_c}{2a}\right)^2 = 0$$

$$\lambda_c = \frac{2a}{m} \quad\quad\quad\quad\quad\quad\quad\quad\quad\quad\quad\quad\quad\quad\quad\quad\quad\quad\quad\quad\quad\quad\quad\quad\quad (4.8)$$

The largest value of cut-off wavelength occurs when $m = 1$.

$$\therefore \lambda_c = 2a.$$

From Equation (4.7), we have:

$$\lambda_p = \frac{\lambda}{\sqrt{1 - \left(\frac{m\lambda}{2a}\right)^2}}.$$

From Equation (4.8), $2a = m\lambda_c$.

$$\therefore \lambda_p = \frac{\lambda}{\sqrt{1 - \left(\frac{\lambda}{\lambda_c}\right)^2}} \quad\quad\quad\quad\quad\quad\quad\quad\quad\quad\quad\quad\quad\quad\quad\quad\quad (4.9)$$

4.9 Cut-off Frequency

The general expression for the cut-off wavelength for TE and TM modes is given by:

$$\lambda_c = \cfrac{2}{\sqrt{\left(\cfrac{m}{a}\right)^2 + \left(\cfrac{n}{b}\right)^2}}, \quad \ldots\ldots(4.10)$$

where a and b are the waveguide dimensions in meters, and

m and n are integers indicating the mode.

m and n cannot both be zero.

The dominant mode occurs when $m = 1$ and $n = 0$.

$v_c = f\lambda = 3 \times 10^8$ m/s.

$$f = \frac{v_c}{\lambda}$$

At cut-off, $\lambda = \lambda_c$ and $f = f_c$.

$$\therefore f_c = \frac{v_c}{\lambda_c}$$

$$= \frac{3 \times 10^8 \sqrt{\left(\cfrac{m}{a}\right)^2 + \left(\cfrac{n}{b}\right)^2}}{2}$$

$$= 1.5 \times 10^8 \sqrt{\left(\frac{m}{a}\right)^2 + \left(\frac{n}{b}\right)^2} \quad \ldots\ldots(4.11)$$

4.10 Group and Phase Velocities

Recall Equation (4.9):

$$\lambda_p = \frac{\lambda}{\sqrt{1-\left(\frac{\lambda}{\lambda_c}\right)^2}}$$

$$v_p = f\lambda_p$$

$$= \frac{f\lambda}{\sqrt{1-\left(\frac{\lambda}{\lambda_c}\right)^2}}$$

$$= \frac{v_c}{\sqrt{1-\left(\frac{\lambda}{\lambda_c}\right)^2}}$$

Hence, phase velocity is given by:

$$v_p = \frac{v_c}{\sqrt{1-\left(\frac{\lambda}{\lambda_c}\right)^2}} \quad \dots\dots\dots\dots\dots\dots\dots\dots\dots\dots\dots\dots\dots\dots\dots\dots\dots\dots\dots\dots\dots\dots (4.12)$$

But $v_p v_g = v_c^2$

$$\therefore v_g = \frac{v_c^2}{v_p}$$

$$= v_c \sqrt{1-\left(\frac{\lambda}{\lambda_c}\right)^2} \quad \dots\dots\dots\dots\dots\dots\dots\dots\dots\dots\dots\dots\dots\dots\dots\dots\dots\dots (4.13)$$

Hence, the velocity of propagation in a waveguide is lower than in free space.

4.11 Impedance Concept

$$Z = \frac{E}{H}$$

$$= \sqrt{\frac{\mu}{\varepsilon}} \quad \text{...(4.14)}$$

$$= \sqrt{\frac{\mu_0 \mu_r}{\varepsilon_0 \varepsilon_r}},$$

where:

E = electric field,

H = magnetic field,

μ = permeability of the medium,

μ_r = relative permeability of the medium,

μ_0 = permeability of free space = $4\pi \times 10^{-7}$ H/m,

ε = permittivity of the medium,

ε_r = relative permittivity of the medium, and

ε_0 = permittivity of free space = $\dfrac{1}{36\pi \times 10^9}$ F/m.

For free space, $\varepsilon_r = 1$ and $\mu_r = 1$.

$$\therefore Z_0 = \sqrt{\frac{\mu_0}{\varepsilon_0}}$$

$$= \sqrt{4\pi \times 10^{-7} \times 36\pi \times 10^9} \text{ ohms}$$

$$= 120\pi \text{ ohms}$$

In a waveguide the characteristic impedance is related to the characteristic impedance of free space as follows:

For TE modes:

$$Z_0' = \frac{Z_0}{\sqrt{1 - \left(\frac{\lambda}{\lambda_c}\right)^2}} \quad \ldots\ldots(4.15)$$

where:

Z_0' = characteristic wave impedance of the waveguide, and

Z_0 = characteristic impedance of free space.

For TM modes:

$$Z_0' = Z_0 \sqrt{1 - \left(\frac{\lambda}{\lambda_c}\right)^2} \quad \ldots\ldots(4.16)$$

4.12 Methods of Exciting Waveguides

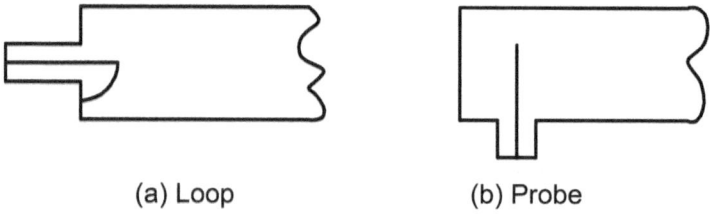

(a) Loop   (b) Probe

Fig. 4.7: Loop and probe coupling of waveguides

When a short antenna in the form of a loop or probe is inserted into a waveguide, it will radiate and if it has been placed correctly the wanted mode will be set up. It is usually placed so as to be parallel to the field which it is desired to set up. The same arrangement may be used at the other end of the waveguide to receive the mode that was set up.

4.13 Circular Waveguides

The behaviour of waves in circular waveguides is the same as in rectangular guides. However, the modes are labeled differently. The formula for cut-off wavelength is different because of the different geometry.

$$\lambda_c = \frac{2\pi r}{k_r}, \quad \dots\dots\dots\dots\dots\dots\dots\dots\dots\dots\dots\dots\dots\dots\dots\dots\dots\dots\dots\dots\dots\dots\dots\dots\dots(4.17)$$

where:

r = internal radius of the waveguide, and

k_r = solution of a Bessel Function Equation [1].

TE Mode	(k_r)	TM Mode	(k_r)
$TE_{0,1}$	3.83	$TM_{0,1}$	2.40
$TE_{1,1}$	1.84	$TM_{1,1}$	3.83
$TE_{2,1}$	3.05	$TM_{2,1}$	5.14
$TE_{0,2}$	7.02	$TM_{0,2}$	5.52
$TE_{1,2}$	5.33	$TM_{1,2}$	7.02
$TE_{2,2}$	6.71	$TM_{2,2}$	8.42

TABLE 4.2: Values of k_r for the principal modes in circular waveguides [1]

4.14 Differences in behaviour between circular and rectangular waveguides

Some differences between circular and rectangular waveguides are:
- Since the mode with the largest cut-off wavelength is the one with the smallest value of (k_r), the $TE_{1,1}$ mode is dominant in circular waveguides. The cut-off wavelength for this mode is given as:

$$\lambda_c = \frac{2\pi r}{1.84}$$
$$= 3.41r$$
$$= 1.7d,$$

where d is the diameter.
- For the circular waveguide m denotes the number of full-wave intensity variations around the circumference and n represents the number of half wave intensity changes radially out from the centre of the wall. [1]

4.15 Advantages of circular waveguides

Circular waveguides have the following advantages:
- They are easier to manufacture compared to rectangular ones.
- They are easier to join together compared to rectangular ones.

4.16 Disadvantage of circular waveguides

The cross-section of a circular waveguide will be much bigger in area than that of a corresponding rectangular waveguide used to carry the same signal. Hence, the space occupied by a circular waveguide system would be considerably more than that for a rectangular system. [1]

4.17 Examples on Waveguides

Example 1
A signal with wavelength of 10cm is to be propagated in a waveguide with air dielectric so that the dominant mode $TE_{1,0}$ wave will propagate with a 25% safety factor. Calculate the breadth of the waveguide and the characteristic wave impedance given the following:

Speed of light in free space $= 3 \times 10^8$ m/s,

Permeability of free space $= 1.257 \times 10^{-6}$ H/m, and

Permittivity of free space $= 8.854 \times 10^{-12}$ F/m.

Solution

$\lambda = 10\,\text{cm} = 0.1\,\text{m}$

$v_c = 3 \times 10^8\,\text{m/s}$

$\mu_0 = 1.257 \times 10^{-6}\,\text{H/m}$

$\varepsilon_0 = 8.854 \times 10^{-12}$ F/m

$$f = \frac{v_c}{\lambda}$$

$$= \frac{3 \times 10^8}{0.1}$$

$$= 3 \times 10^9 \text{ Hz.}$$

But $f = 1.25 f_c$.

$$\therefore f_c = \frac{f}{1.25}$$

$$= \frac{3 \times 10^9}{1.25}$$

$$= 2.4 \times 10^9 \text{ Hz}$$

$$= 2.4 \text{GHz}$$

$$f_c = 1.5 \times 10^8 \sqrt{\left(\frac{m}{a}\right)^2 + \left(\frac{n}{b}\right)^2}$$

For the $TE_{1,0}$ mode, $m = 1$ and $n = 0$.

$$\therefore f_c = 1.5 \times 10^8 \sqrt{\left(\frac{1}{a}\right)^2}$$

$$= \frac{1.5 \times 10^8}{a}$$

$$\therefore a = \frac{1.5 \times 10^8}{f_c}$$

$$= \frac{1.5 \times 10^8}{2.4 \times 10^9}$$

$= 0.0625 \text{m}$

$= 6.25 \text{cm}.$

$$Z_0 = \sqrt{\frac{\mu_0}{\varepsilon_0}}$$

$$= \sqrt{\frac{1.257 \times 10^{-6}}{8.854 \times 10^{-12}}}$$

$$= \sqrt{0.142 \times 10^6}$$

$= 377 \text{ ohms}.$

$$\lambda_c = \frac{2}{\sqrt{\left(\frac{m}{a}\right)^2 + \left(\frac{n}{b}\right)^2}}$$

$$= \frac{2}{\sqrt{\left(\frac{1}{a}\right)^2 + \left(\frac{0}{b}\right)^2}}$$

$= 2a$

$\therefore \lambda_c = 2 \times 0.0625 = 0.125 \text{ m}.$

$$Z_0' = \frac{Z_0}{\sqrt{1 - \left(\frac{\lambda}{\lambda_c}\right)^2}}$$

$$= \frac{377}{\sqrt{1-\left(\frac{0.1}{0.125}\right)^2}}$$

$= 628.3$ ohms

Example 2

A signal of frequency 6 GHz is propagated in a rectangular waveguide whose internal width is 5.0 cm. Calculate, for the dominant mode $TE_{1,0}$, :

a) The cut-off wavelength,
b) The guide wavelength,
c) The group and phase velocities, and
d) The characteristic wave impedance.

Solution

a) $\lambda_c = \dfrac{2}{\sqrt{\left(\dfrac{m}{a}\right)^2 + \left(\dfrac{n}{b}\right)^2}}$

For the $TE_{1,0}$ mode, $m = 1$ and $n = 0$.

$\therefore \lambda_c = \dfrac{2}{\sqrt{\left(\dfrac{1}{a}\right)^2 + \left(\dfrac{0}{b}\right)^2}}$

$= 2a$

$= 2 \times 5 = 10\,\text{cm}.$

b) The free space wavelength λ is given by:

$$\lambda = \frac{v_c}{f}$$

$$= \frac{3 \times 10^8}{6 \times 10^9}$$

$$= 0.05 \text{m}$$

$$= 5 \text{cm}$$

The guide wavelength is the phase wavelength given by:

$$\lambda_p = \frac{\lambda}{\sqrt{1-\left(\frac{\lambda}{\lambda_c}\right)^2}}$$

$$= \frac{5}{\sqrt{1-\left(\frac{5}{10}\right)^2}}$$

$$= \frac{5}{\sqrt{1-0.25}}$$

$$= 0.0577 \text{m}$$

$$= 5.77 \text{ cm.}$$

c) The group velocity is given by:

$$v_g = v_c \sqrt{1-\left(\frac{\lambda}{\lambda_c}\right)^2}$$

$$= 3 \times 10^8 \sqrt{1 - \left(\frac{5}{10}\right)^2}$$

$$= 3 \times 10^8 \times 0.866$$

$$= 2.6 \times 10^8 \text{ m/s}$$

The phase velocity is given by:

$$v_p = \frac{v_c}{\sqrt{1 - \left(\frac{\lambda}{\lambda_c}\right)^2}}$$

$$= \frac{3 \times 10^8}{\sqrt{1 - \left(\frac{5}{10}\right)^2}}$$

$$= \frac{3 \times 10^8}{0.866}$$

$$= 3.46 \times 10^8 \text{ m/s}$$

d) The characteristic wave impedance is given by:

$$Z_0' = \frac{Z_0}{\sqrt{1 - \left(\frac{\lambda}{\lambda_c}\right)^2}}$$

$$= \frac{120\pi}{\sqrt{1 - \left(\frac{5}{10}\right)^2}}$$

$$= \frac{120\pi}{0.866}$$

$$= 435.3 \text{ ohms}$$

Example 3

A rectangular waveguide is 5.5 cm x 2.5 cm – inside measurement. Calculate:

a) The cut-off frequency of the dominant mode.

b) The lowest frequency and the mode closest to the dominant mode.

Solution

a) The dominant mode in a rectangular waveguide is the $TE_{1,0}$ mode with $m = 1$ and $n = 0$.

$$f_c = 1.5 \times 10^8 \sqrt{\left(\frac{m}{a}\right)^2 + \left(\frac{n}{b}\right)^2}$$

$$= 1.5 \times 10^8 \sqrt{\left(\frac{1}{a}\right)^2}$$

$$= \frac{1.5 \times 10^8}{a}$$

$$= \frac{1.5 \times 10^8}{0.055}$$

$$= 2.73 \text{ GHz}$$

b) The possible TE modes after the dominant mode are: $TE_{0,1}$, $TE_{2,0}$, and $TE_{0,2}$.

The different cut-off frequencies are:

For $TE_{0,1}$:

$$f_c = 1.5 \times 10^8 \sqrt{\left(\frac{0}{0.055}\right)^2 + \left(\frac{1}{0.025}\right)^2}$$

$$= 1.5 \times 10^8 \sqrt{\left(\frac{1}{0.025}\right)^2}$$

$$= \frac{1.5 \times 10^8}{0.025} \text{ Hz}$$

$$= 6.0 \text{ GHz}$$

For $TE_{2,0}$:

$$f_c = 1.5 \times 10^8 \sqrt{\left(\frac{2}{0.055}\right)^2 + \left(\frac{0}{0.025}\right)^2}$$

$$= 1.5 \times 10^8 \sqrt{\left(\frac{2}{0.055}\right)^2}$$

$$= \frac{1.5 \times 10^8 \times 2}{0.055} \text{ Hz}$$

$$= 5.45 \text{ GHz}$$

For $TE_{0,2}$:

$$f_c = 1.5 \times 10^8 \sqrt{\left(\frac{0}{0.055}\right)^2 + \left(\frac{2}{0.025}\right)^2}$$

$$= 1.5 \times 10^8 \sqrt{\left(\frac{2}{0.025}\right)^2}$$

$$= \frac{1.5 \times 10^8 \times 2}{0.025} \text{ Hz}$$

$$= 12.0 \text{ GHz}$$

Hence, the mode closest to the dominant mode is the $TE_{2,0}$ with a cut-off frequency of 5.45 GHz.

Example 4

It is necessary to propagate a 10-GHz signal in a waveguide whose wall separation is 6 cm.

a) What is the greatest number of half-waves of electric intensity which it will be possible to establish between the two walls?

b) Calculate the guide wavelength for this mode of propagation.

Solution

a) The free wavelength is given by:

$$\lambda = \frac{v_c}{f}$$

$$= \frac{3 \times 10^8}{10 \times 10^9}$$

$$= 0.03 \text{ m}$$

= 3 cm

The wave will propagate in the waveguide as long as the waveguide's cut-off wavelength is greater than the free space wavelength of the signal.

It is necessary to calculate the cut-off wavelengths of the guide for increasing values of m.

The cut-off wavelength is given by:

$$\lambda_c = \frac{2}{\sqrt{\left(\frac{m}{a}\right)^2 + \left(\frac{n}{b}\right)^2}}$$

When $n = 0$, $\lambda_c = \frac{2a}{m}$.

But $a = 6$ cm.

$$\therefore \lambda_c = \frac{12}{m} \text{ cm.}$$

When $m = 1$, $\lambda_c = 12$ cm $> \lambda$. Hence, the mode will propagate.

When $m = 2$, $\lambda_c = 6$ cm $> \lambda$. Hence, the mode will propagate.

When $m = 3$, $\lambda_c = 4$ cm $> \lambda$. Hence, the mode will propagate.

When $m = 4$, $\lambda_c = 3$ cm $= \lambda$. Hence, the mode will not propagate.

Hence, the greatest number of half-waves of electric intensity that can be established between the walls is 3.

b) When $m = 3$, $\lambda_c = 4$ cm.

Therefore, the guide wavelength is given by:

Waveguides

$$\lambda_p = \frac{\lambda}{\sqrt{1-\left(\frac{\lambda}{\lambda_c}\right)^2}}$$

$$= \frac{3}{\sqrt{1-\left(\frac{3}{4}\right)^2}}$$

$$= \frac{3}{0.6614}$$

$$= 0.0454 \text{m}$$

$$= 4.54 \text{ cm}.$$

Example 5

A signal of frequency 10 GHz is propagated in a circular waveguide of internal diameter 5 cm. For the $TE_{1,1}$ mode calculate:

a) The cut-off wavelength.
b) The guide wavelength.
c) The characteristic wave impedance.

Solution

a) $\lambda = \dfrac{v_c}{f}$

$$= \frac{3 \times 10^8}{10 \times 10^9}$$

$= 0.03\text{m}$

$= 3\text{cm}$

$\lambda_c = \dfrac{2\pi r}{k_r}$

$r = 2.5\,\text{cm}$

For the $TE_{1,1}$ mode, $k_r = 1.84$.

$\therefore \lambda_c = \dfrac{2\pi \times 2.5}{1.84}$

$= 8.54\,\text{cm}.$

b) The guide wavelength is given by:

$$\lambda_p = \dfrac{\lambda}{\sqrt{1-\left(\dfrac{\lambda}{\lambda_c}\right)^2}}$$

$$= \dfrac{3}{\sqrt{1-\left(\dfrac{3}{8.54}\right)^2}}$$

$$= \dfrac{3}{0.9363}$$

$= 0.032\text{m}$

$= 3.2\,\text{cm}.$

c) The characteristic wave impedance is given by:

$$Z_0' = \frac{Z_0}{\sqrt{1-\left(\dfrac{\lambda}{\lambda_c}\right)^2}}$$

$$= \frac{120\pi}{\sqrt{1-\left(\dfrac{3}{8.54}\right)^2}}$$

$$= \frac{120\pi}{0.9363}$$

$$= 403 \text{ ohms}$$

Example 6

A 6-GHz signal is to be propagated in the dominant mode in a rectangular waveguide. If its group velocity is to be 80% of the free-space velocity of light,

    a) What must be the breath of the waveguide?

    b) What impedance will it offer to this signal if it is correctly matched?

Solution

    a) The free wavelength is given by:

$$\lambda = \frac{v_c}{f}$$

$$= \frac{3\times 10^8}{6\times 10^9}$$

$$= 0.05 \text{m}$$
$$= 5 \text{cm}$$
$$v_g = 80\% v_c$$
$$= 0.8 v_c$$
$$= 0.8 \times 3 \times 10^8$$
$$= 2.4 \times 10^8 \text{ m/s}$$

$$v_g = v_c \sqrt{1 - \left(\frac{\lambda}{\lambda_c}\right)^2}$$

$$\therefore 2.4 \times 10^8 = 3 \times 10^8 \sqrt{1 - \left(\frac{\lambda}{\lambda_c}\right)^2}$$

$$\sqrt{1 - \left(\frac{\lambda}{\lambda_c}\right)^2} = 0.8$$

$$1 - \left(\frac{\lambda}{\lambda_c}\right)^2 = 0.64$$

$$\left(\frac{\lambda}{\lambda_c}\right)^2 = 0.36$$

$$\frac{\lambda}{\lambda_c} = 0.6$$

$$\lambda_c = \frac{\lambda}{0.6}$$

But $\lambda = 5$ cm.

$$\therefore \lambda_c = \frac{5}{0.6} = 8.33 \text{ cm}.$$

$$\lambda_c = \frac{2}{\sqrt{\left(\dfrac{m}{a}\right)^2 + \left(\dfrac{n}{b}\right)^2}}$$

For the $TE_{1,0}$ mode, $m = 1$ and $n = 0$.

$$\lambda_c = \frac{2}{\sqrt{\left(\dfrac{1}{a}\right)^2 + \left(\dfrac{0}{b}\right)^2}}$$

$\therefore \lambda_c = 2a$

$a = \dfrac{\lambda_c}{2} = \dfrac{8.33}{2}$

$\quad = 4.2\,\text{cm}$

b) $Z_0' = \dfrac{Z_0}{\sqrt{1 - \left(\dfrac{\lambda}{\lambda_c}\right)^2}}$

$\quad = \dfrac{120\pi}{0.8}$

$\quad = 471\text{ ohms}$

Example 7

A 10-GHz signal is to be propagated in a waveguide whose internal dimensions are 7.5cm x 3.75cm. Calculate the characteristic wave impedance of this rectangular waveguide for the first three $TE_{m,0}$ modes and for the $TM_{1,1}$ mode.

Solution

$f = 10\,\text{GHz}$

$a = 7.5\,\text{cm}$

$b = 3.75\,\text{cm}$

$v_c = 3 \times 10^8 \,\text{m/s}$

$\lambda = \dfrac{v_c}{f}$

$= \dfrac{3 \times 10^8}{10 \times 10^9}$

$= 0.03\,\text{m}$

$= 3\,\text{cm}$

$\lambda_c = \dfrac{2}{\sqrt{\left(\dfrac{m}{a}\right)^2 + \left(\dfrac{n}{b}\right)^2}}$

For the $TE_{m,0}$ modes, $n = 0$.

$\lambda_c = \dfrac{2}{\sqrt{\left(\dfrac{m}{a}\right)^2 + \left(\dfrac{0}{b}\right)^2}}$

$$= \frac{2a}{m}$$

Since $a = 7.5$ cm, $\lambda_c = \frac{15}{m}$ cm.

$$Z_0' = \frac{Z_0}{\sqrt{1-\left(\frac{\lambda}{\lambda_c}\right)^2}}$$

$$= \frac{120\pi}{\sqrt{1-\left(\frac{\lambda}{\lambda_c}\right)^2}} \text{ ohms}$$

For the $TE_{1,0}$ mode, $m = 1$ so $\lambda_{c_1} = 15$ cm.

$$Z_{01}' = \frac{120\pi}{\sqrt{1-\left(\frac{3}{15}\right)^2}} \text{ ohms}$$

$$= 384.8 \text{ ohms}$$

For the $TE_{2,0}$ mode, $m = 2$ so $\lambda_{c_2} = 7.5$ cm.

$$Z_{02}' = \frac{120\pi}{\sqrt{1-\left(\frac{3}{7.5}\right)^2}} \text{ ohms}$$

$$= 411.3 \text{ ohms}$$

For the $TE_{3,0}$ mode, $m = 3$ so $\lambda_{c_3} = 5$ cm.

$$Z_{03}' = \frac{120\pi}{\sqrt{1-\left(\frac{3}{5}\right)^2}} \text{ ohms}$$

= 471.2 ohms

For the $TM_{1,1}$ mode, $m = 1$ and $n = 1$.

$$\lambda_{c_4} = \frac{2}{\sqrt{\left(\frac{1}{7.5}\right)^2 + \left(\frac{1}{3.75}\right)^2}}$$

$$= \frac{2}{\sqrt{0.01778 + 0.07111}}$$

$$= \frac{2}{0.298146}$$

= 6.71 cm.

$$Z_{04}' = 120\pi \sqrt{1 - \left(\frac{\lambda}{\lambda_c}\right)^2}$$

$$= 120\pi \sqrt{1 - \left(\frac{3}{6.71}\right)^2}$$

= 337.2 ohms.

4.18 Questions on Waveguides

Question 1

Design a rectangular waveguide with copper conductor and air dielectric for a signal of wavelength 6cm so that the $TE_{1,0}$ wave will propagate with a 30% safety factor $(f = 1.3 f_c)$ Calculate the characteristic wave impedance given the following:

Speed of light in free space $= 3 \times 10^8$ m/s,

Permeability of free space $= 1.257 \times 10^{-6}$ H/m, and

Permittivity of free space $= 8.854 \times 10^{-12}$ F/m.

Answer: $a = 3.9$ cm; $b = 1.95$ cm; $Z_0' = 590$ ohms

Question 2

It is required to propagate a 12 GHz signal in a rectangular waveguide in such a manner that the characteristic wave impedance is 450 ohms. If the $TE_{1,0}$ mode is used, what must be the corresponding cross-sectional wave dimensions? If the guide is 60 cm long how long will this signal take to travel from one end of the waveguide to the other?

Answer: $a = 3.9$ cm; $b = 1.95$ cm; $t = 2.4$ns

Question 3

A waveguide has an internal width of 4 cm and carries the dominant mode of a signal of unknown frequency. If the characteristic wave impedance is 600 ohms, calculate the frequency of the signal.

Answer: 4.8 GHz

Question 4

A rectangular waveguide measures 5.0 x 3.0 cm internally and has a 10-GHz signal propagated in it. Calculate:

   a) The cut-off wavelength.
   b) The guide wavelength.
   c) The group and phase velocities.
   d) The characteristic wave impedance.

Answer: a) 5.15 cm b) 3.7 cm c) 2.44×10^8 m/s; 3.69×10^8 m/s d) 306 ohms.

Question 5

Calculate the ratio of the cross-section of a circular waveguide to that of a rectangular one if each is to have the same cut-off wavelength for its dominant mode.

Answer: 2.17

5

Antennas

5.1 Definition

An antenna is a structure that is generally a metallic object, often a wire or group of wires, used to convert high frequency current into electromagnetic waves and vice versa. Transmitting and receiving antennas have similar characteristics. The spacing, length, and shape of the device are related to the wavelength, λ, of the desired frequency. [1]

5.2 Radiation from a current element in free space

The Elementary Doublet is a theoretical antenna shorter than a wavelength. It is used as a standard to which all other antenna characteristics are compared. [1]

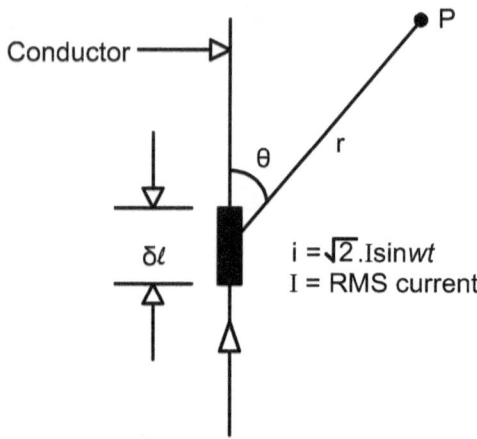

Fig. 5.1: Radiation from a current element in free space – elementary doublet (Hertzian dipole)

The magnetizing force at point P due to the current element δl is given by:

$$\delta H = \frac{\sqrt{2} \cdot I \delta l \sin \theta}{4\pi} \left[\frac{1}{r^2} \sin w\left(t - \frac{r}{v}\right) + \frac{w}{rv} \cos w\left(t - \frac{r}{v}\right) \right] \quad \ldots (5.1)$$

The component $\dfrac{\sqrt{2} \cdot I \delta l}{4\pi r^2} \sin \theta \sin w\left(t - \dfrac{r}{v}\right)$ represents the Induction field.

The second component $\dfrac{w\sqrt{2} \cdot I \delta l}{4\pi r v} \sin \theta \cos w\left(t - \dfrac{r}{v}\right)$ represents the Radiation field.

Since the Induction field is inversely proportional to r^2 it will predominate at points close to the current element where r is small. At great distances where r is large the Induction field will be negligible

Antennas

compared to the Radiation field. The two fields will have equal amplitude at the critical value of r where:

$$\frac{1}{r_c^2} = \frac{w}{r_c v}.$$

$$\therefore r_c = \frac{v}{w}.$$

In free space, $v = c$, the speed of light.

Thus $v = \lambda f$.

$$\therefore r_c = \frac{\lambda f}{2\pi f}$$

$$= \frac{\lambda}{2\pi}.$$

The Radiation field contributes to a flow of energy away from the antenna whereas the Induction field contributes to energy that is stored in the field during one quarter of a cycle and returned to the circuit during the next quarter. Considering only the Radiation field, we have:

$$\delta H = \frac{w\sqrt{2} \cdot I\delta l}{4\pi r v} \sin\theta \cos w\left(t - \frac{r}{v}\right). \quad\quad\quad (5.2)$$

$$\frac{w}{rv} = \frac{2\pi f}{r\lambda f} = \frac{2\pi}{r\lambda}.$$

$$\therefore \delta H = \frac{\sqrt{2}}{2r\lambda} I\delta l \sin\theta \cos w\left(t - \frac{r}{v}\right) \quad\quad\quad (5.3)$$

Note that $\frac{\delta l}{\lambda}$ is the length of the antenna in wavelengths.

5.3 Power radiated by a doublet

$P_F = E \times H$,

where:

P_F = Power flow in Watts/metre^2,

E = Electrical intensity in Volts/metre, and

H = Magnetic intensity in Amps/metre.

But $E = Z_0 H$,

where $Z_0 = 120\pi$ is the characteristic impedance of free space.

$\therefore P_F = 120\pi H^2$.

The instantaneous power flow is given by:

$P_{F_i} = 120\pi \delta H^2$

Substituting δH from Equation (5.3) yields:

$P_{F_i} = 120\pi \left(\dfrac{\sqrt{2} \cdot I\delta l \sin\theta}{2r\lambda} \right)^2 \cos^2 w\left(t - \dfrac{r}{v}\right)$.

The average value of $\cos^2 w\left(t - \dfrac{r}{v}\right) = \dfrac{1}{2}$.

Hence, the average value of the power flow is given by:

$P_{F_{av}} = 120\pi \left(\dfrac{\sqrt{2} \cdot I\delta l \sin\theta}{2r\lambda} \right)^2 \times \dfrac{1}{2}$

$= 120\pi \left(\dfrac{I\delta l \sin\theta}{2r\lambda} \right)^2$ ....(5.4)

The total power radiated is the average power flow in all directions. The antenna is considered to be the centre of a sphere.

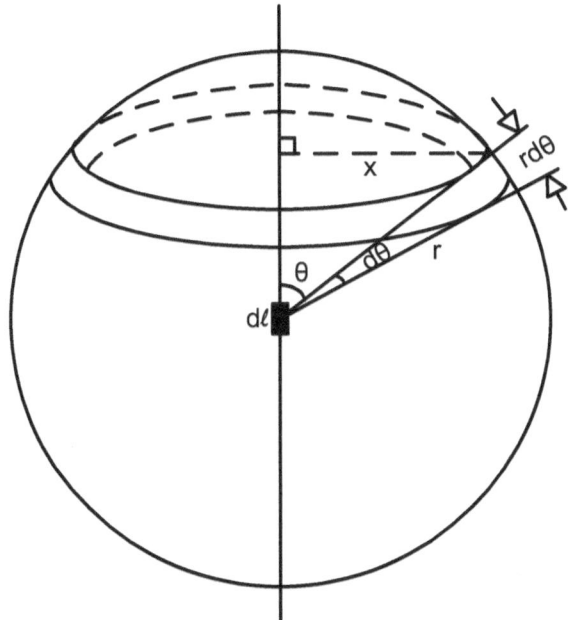

Fig. 5.2: Total power radiated by a doublet

The radius of a strip is $x = r\sin\theta$.

The area of a strip is given by:

$2\pi r\sin\theta \times rd\theta = 2\pi r^2 \sin\theta d\theta$.

Power flow through the strip is given by: Area $\times P_{F_{av}}$.

$$\therefore \delta P = 2\pi r^2 \sin\theta d\theta \times 120\pi \left(\frac{I\delta l \sin\theta}{2r\lambda}\right)^2$$

$$= 60\pi^2 I^2 \left(\frac{\delta l}{\lambda}\right)^2 \sin^3\theta d\theta$$

$$\therefore P = \int_0^\pi 60\pi^2 I^2 \left(\frac{\delta l}{\lambda}\right)^2 \sin^3\theta d\theta$$

$$P = 60\pi^2 I^2 \left(\frac{\delta l}{\lambda}\right)^2 \int_0^{2\pi} \sin^3\theta d\theta$$

$$= 60\pi^2 I^2 \left(\frac{\delta l}{\lambda}\right)^2 \times \frac{4}{3}$$

$$= 80\pi^2 I^2 \left(\frac{\delta l}{\lambda}\right)^2 \text{ Watts} \quad\quad\quad\quad\quad\quad\quad\quad\quad\quad\quad\quad (5.5)$$

P is the total radiated power. It can also be represented as:

$P = I^2 R_r$ Watts,

where $R_r = 80\pi^2 \left(\frac{\delta l}{\lambda}\right)^2$ ohms.

R_r is the radiation resistance.

The radiation resistance is defined as the resistance which will absorb the same power from the same current as that current actually radiated. It represents the resistive part of the impedance of the antenna as seen by the source of power. Any losses in the antenna will add to this resistance.

5.4 Antenna Losses and Efficiency

In addition to the energy radiated by an antenna there are power losses caused by the following:
- Ground resistance

Antennas

- Eddy currents induced in nearby metallic objects.
- I^2R (Heating) losses in the antenna.

$$P_{in} = P_d + P_r.$$

where:

P_{in} = Power delivered to the feed point (input power),

P_d = Power dissipated (lost), and

P_r = Power radiated (useful power).

$$I^2 R_{in} = I^2 R_d + I^2 R_r.$$

$$\therefore R_{in} = R_d + R_r.$$

Antenna efficiency can be defined as:

$$\eta = \frac{P_r}{P_{in}}.$$

$$\eta = \frac{I^2 R_r}{I^2 R_{in}}$$

$$= \frac{R_r}{R_{in}}$$

$$= \frac{R_r}{R_r + R_d} \times 100\% \quad \ldots\ldots(5.7)$$

5.5 Antenna Gain

Gain in any given direction is defined as:

$$\text{Gain} = \frac{P_F}{P_A},$$

where:

P_F = Power flow in the direction, and

P_A = Average power flow = $\frac{P_{in}}{4\pi r^2}$.

Hence, $\text{Gain} = \frac{4\pi r^2 P_F}{P_{in}}$,

where P_{in} = Input power.

5.6 Gain of a Doublet

From Equation (5.4),

$$P_F = 120\pi \left(\frac{I\delta l \sin\theta}{2r\lambda} \right)^2.$$

From Equation (5.5), the total radiated power is given by:

$$P = 80\pi^2 I^2 \left(\frac{\delta l}{\lambda} \right)^2.$$

Hence, $\text{Gain} = 4\pi r^2 \times 120\pi \left(\frac{I\delta l \sin\theta}{2r\lambda} \right)^2 \times \frac{\lambda^2}{80\pi^2 I^2 (\delta l)^2}.$

$\therefore \text{Gain} = 1.5 \sin^2\theta$ ....(5.8)

Maximum gain occurs when $\theta = 90°$.

∴ Maximum Gain = 1.5.

5.7 Polar Diagram

This is a graphical presentation of the field strength in magnitude and direction.

From Equation (5.3), we have:

$$\delta H = \frac{\sqrt{2}}{2r\lambda} I \delta l \sin \theta \cos w\left(t - \frac{r}{v}\right).$$

Considering only the magnitude:

$$\delta H = \frac{\sqrt{2}}{2r\lambda} I \delta l \sin \theta.$$

Fig. 5.3 represents the magnitude of the radiation field in the different directions given in TABLE 5.1.

θ (degrees)	Radiation Field δH
0	0
45	$\dfrac{I\delta l}{2r\lambda}$
90	$\dfrac{\sqrt{2}\cdot I\delta l}{2r\lambda}$
135	$\dfrac{I\delta l}{2r\lambda}$
180	0
225	$-\dfrac{I\delta l}{2r\lambda}$
270	$-\dfrac{\sqrt{2}\cdot I\delta l}{2r\lambda}$
315	$-\dfrac{I\delta l}{2r\lambda}$
360	0

TABLE 5.1: Values of radiation field in different directions

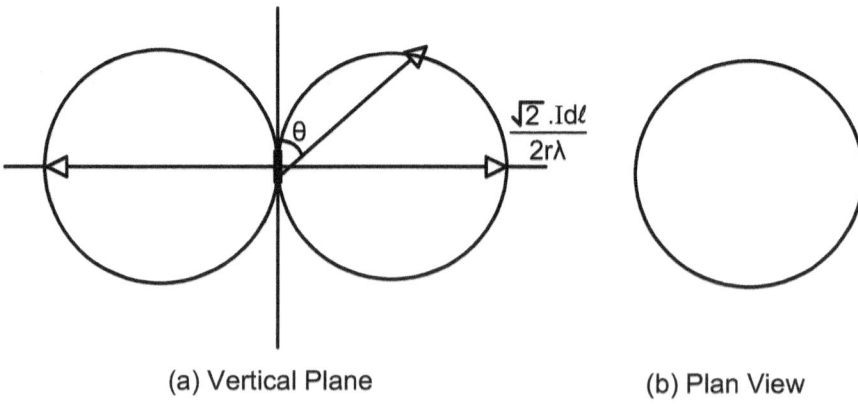

(a) Vertical Plane   (b) Plan View

Fig. 5.3: Polar diagram of an elementary doublet

5.8 Power radiated by a short dipole

From Equation (5.5), the power radiated by a current element is given by:

$$P = 80\pi^2 I^2 \left(\frac{\delta l}{\lambda}\right)^2 \text{ Watts.}$$

This assumes that the amplitude of the current element does not change over the length of the element.

For the dipole $l > \delta l$ and the amplitude of the current will change over the length as shown in Fig. 5.4. Hence, the power radiated by the dipole will be less than that produced by the current element.

This fact will be taken into account by using an equivalent length l_e for the dipole.

$$\therefore P = 80\pi^2 I^2 \left(\frac{l_e}{\lambda}\right)^2 \text{ Watts.} \quad \text{............(5.9)}$$

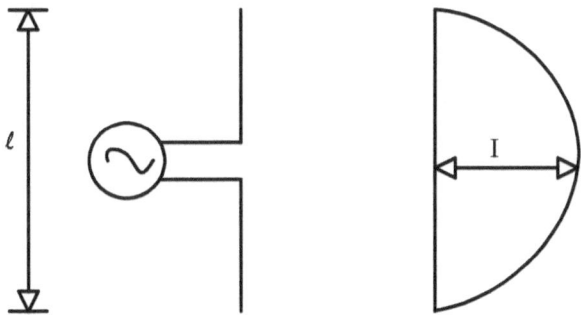

Fig. 5.4: Current distribution in a short dipole

5.9 Power radiated by a half-wavelength dipole in free space

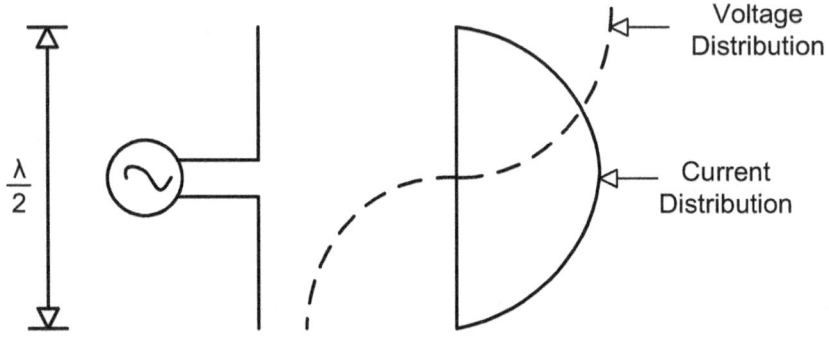

Fig. 5.5: Current and voltage distributions in a half-wavelength dipole

The current and voltage distributions are almost the same as for a loss-free open-circuited quarter wavelength transmission line. The actual distribution is modified slightly because the antenna radiates and is therefore not loss-free. If the current is assumed to be cosinusoidally distributed from the centre of the dipole and the phase difference is ignored, the equivalent length l_e is given by:

$$l_e = \frac{l}{\pi} \int_{-\frac{\pi}{2}}^{\frac{\pi}{2}} \cos\theta \, d\theta$$

$$= \frac{l}{\pi} [\sin\theta]_{-0.5\pi}^{0.5\pi}$$

$$= \frac{2l}{\pi}.$$

$$P = 80\pi^2 I^2 \left(\frac{l_e}{\lambda}\right)^2$$

$$= 80\pi^2 I^2 \left(\frac{2l}{\pi\lambda}\right)^2$$

$$= 320 I^2 \left(\frac{l}{\lambda}\right)^2.$$

But $l = \frac{\lambda}{2}$ for a half-wavelength dipole.

$$\therefore P = 320 I^2 \left(\frac{1}{2}\right)^2$$

$$= 80 I^2 \text{ Watts.} \quad\quad\quad\quad\quad\quad\quad\quad\quad\quad\quad\quad (5.10)$$

Hence, the radiation resistance for a half-wavelength dipole is given by:

$R_r = 80$ ohms.

Equation (5.10) is a good approximation for a half-wavelength dipole. A more exact value is given by:

$$P = 73 I^2 \text{ Watts. [5]} \quad\quad\quad\quad\quad\quad\quad\quad\quad\quad (5.11)$$

5.10 Power radiated by a short vertical earthed antenna

Fig. 5.6 depicts a vertical earthed antenna and its radiated image.

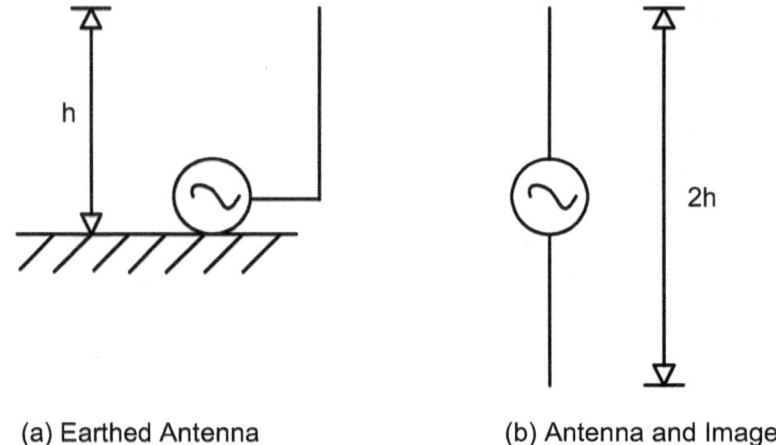

(a) Earthed Antenna   (b) Antenna and Image

Fig. 5.6: Vertical earthed antenna

If the height of the antenna is short ($h < \frac{\lambda}{10}$), then the radiated power will be given by:

$$P = 80\pi^2 I^2 \left(\frac{2h_e}{\lambda}\right)^2 \text{ Watts,}$$

where h_e = equivalent height of the antenna.

$$\therefore P = 320\pi^2 I^2 \left(\frac{h_e}{\lambda}\right)^2 \text{ Watts.}$$

Since there is no field below the earth, the power radiated is half that of the antenna and its image.

$$\therefore P = 160\pi^2 I^2 \left(\frac{h_e}{\lambda}\right)^2 \text{ Watts} \dots\dots\dots\dots\dots\dots\dots\dots\dots\dots\dots\dots\dots\dots\dots\dots(5.12)$$

$$h_e = \frac{2h}{\pi}.$$

$$\therefore P = 160\pi^2 I^2 \left(\frac{2h}{\pi\lambda}\right)^2$$

$$= 640 \left(\frac{h}{\lambda}\right)^2 I^2$$

Assume $h = \frac{\lambda}{4}$.

$$\therefore P = 640 \left(\frac{\lambda}{4\lambda}\right)^2 I^2$$

$$= 40 I^2 \text{ Watts } \dots\dots\dots\dots\dots\dots\dots\dots\dots\dots\dots\dots\dots\dots\dots\dots\dots\dots\dots\dots(5.13)$$

A more exact value is given by:

$P = 36.5 I^2$ Watts. [5]$\dots\dots\dots\dots\dots\dots\dots\dots\dots\dots\dots\dots\dots\dots\dots\dots\dots\dots(5.14)$

5.11 Effect of the earth on antennas

Part of the radiation from any antenna will be directed towards the earth or sea which will give rise to reflection. In Fig. 5.7, the total field at point P is the sum of the direct wave and the reflected wave.

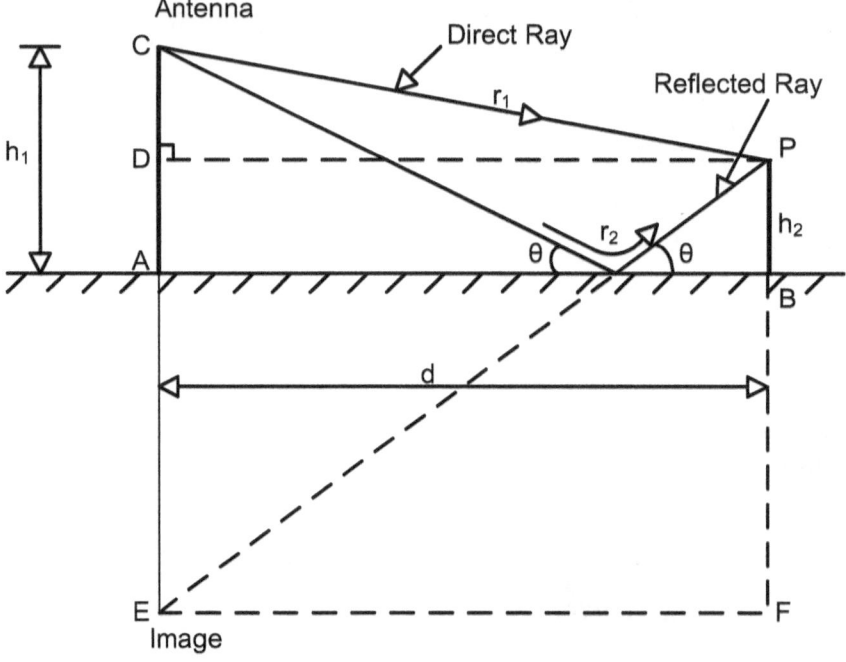

Fig. 5.7: Direct and reflected rays from an antenna

Assume that the earth is flat and a perfect conductor. The reflected wave can be regarded as originating from an image of the antenna below the surface of the earth. The total field at P can be regarded as that due to two antennas with the appropriate phase difference. [5]

From $\triangle CDP$,

$$r_1^2 = d^2 + (h_1 - h_2)^2$$
$$= d^2 \left[1 + \frac{(h_1 - h_2)^2}{d^2} \right]$$

$$\therefore r_1 = d\sqrt{\left[1+\frac{(h_1-h_2)^2}{d^2}\right]}$$

If $d >> (h_1 - h_2)$, then:

$$r_1 = d\left[1+\frac{(h_1-h_2)^2}{2d^2}\right].$$

From $\triangle PEF$,

$$r_2^2 = d^2 + (h_1+h_2)^2$$

$$= d^2\left[1+\frac{(h_1+h_2)^2}{d^2}\right]$$

$$\therefore r_2 = d\sqrt{\left[1+\frac{(h_1+h_2)^2}{d^2}\right]}$$

If $d >> (h_1 + h_2)$, then:

$$r_2 = d\left[1+\frac{(h_1+h_2)^2}{2d^2}\right].$$

If the path difference between the reflected ray and the direct ray is x, then:

$$x = r_2 - r_1$$

$$= d\left[1+\frac{(h_1+h_2)^2}{2d^2}\right] - d\left[1+\frac{(h_1-h_2)^2}{2d^2}\right]$$

$$= \frac{2h_1 h_2}{d}$$

$\lambda = 2\pi$

$$\therefore x = \frac{2\pi}{\lambda} \times \frac{2h_1 h_2}{d}$$

$$= \frac{4\pi h_1 h_2}{d\lambda} \text{ radians.}$$

Hence, the phase difference between the two fields is given by:

$\frac{4\pi h_1 h_2}{d\lambda}$ radians.

$$= \frac{720 h_1 h_2}{d\lambda} \text{ degrees. [5]} \quad\quad\quad\quad\quad\quad\quad\quad\quad\quad\quad\quad\quad (5.14)$$

5.12 General theory of antenna array

Let the array consist of N identical elements each carrying the same current and each having a circular polar diagram as described in Section 5.7.

Let the elements be equally spaced by d meters, as shown in Fig. 5.8.

Let the phase of the current be advanced by ϕ in a direction $\theta = 0$.

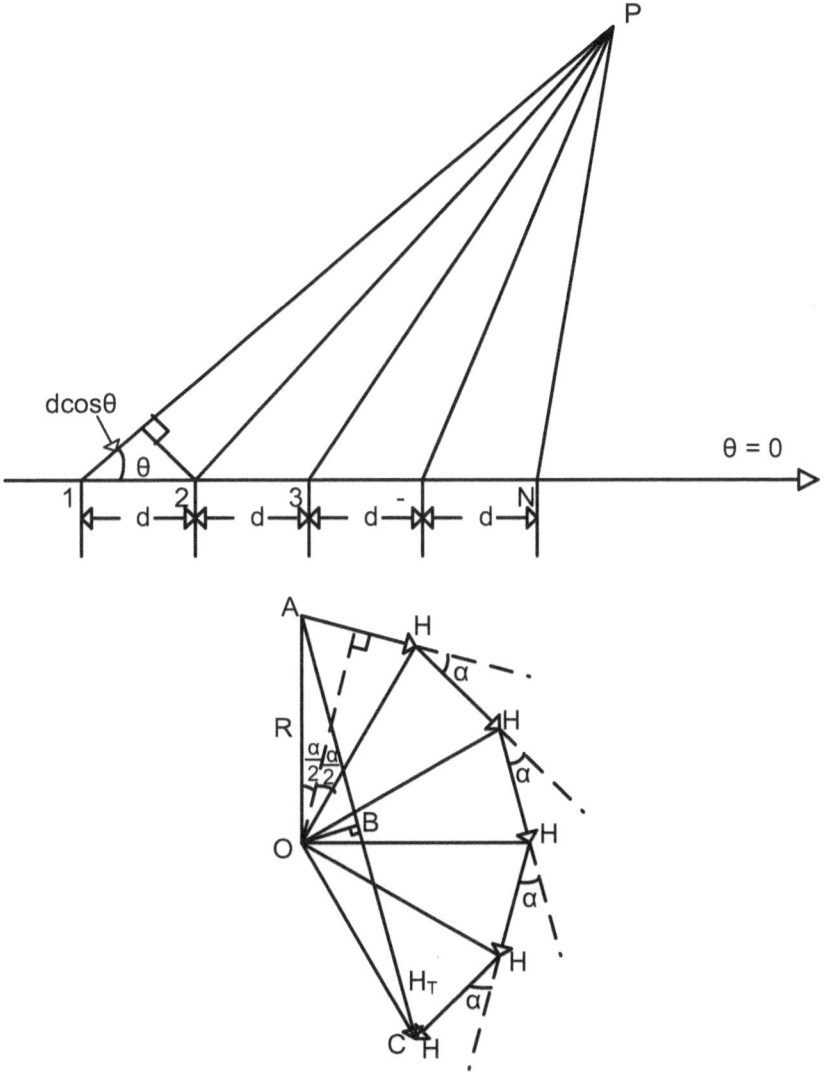

Fig. 5.8: Antenna arrays with fields

The phase of the field at P due to two adjacent elements differs by an angle α given by:

$$\alpha = d\cos\theta \cdot \frac{2\pi}{\lambda} + \phi$$

$$= \frac{2\pi d \cos\theta}{\lambda} + \phi \quad\quad\quad\quad\quad\quad\quad\quad\quad\quad\quad\quad\quad\quad\quad (5.15)$$

The total magnetic field at P, H_T, is the vector sum of N equal fields differing by an angle α.

For each magnetic field,

$$\frac{H}{2} = R\sin\left(\frac{\alpha}{2}\right)$$

$$\therefore R = \frac{H}{2\sin\left(\frac{\alpha}{2}\right)}$$

For the total field at P,

$$\frac{H_T}{2} = R\sin\left(\frac{N\alpha}{2}\right).$$

$$\therefore H_T = 2R\sin\left(\frac{N\alpha}{2}\right).$$

$$= \frac{2\sin\left(\frac{N\alpha}{2}\right)}{1} \times \frac{H}{2\sin\left(\frac{\alpha}{2}\right)}$$

$$= \frac{H\sin\left(\frac{N\alpha}{2}\right)}{\sin\left(\frac{\alpha}{2}\right)}$$

$$= \frac{NH \sin\left(\frac{N\alpha}{2}\right)}{N \sin\left(\frac{\alpha}{2}\right)}$$

$$= NH \left[\frac{\sin\left(\frac{N\alpha}{2}\right)}{N \sin\left(\frac{\alpha}{2}\right)} \right] \quad \text{...........(5.16)}$$

But $\alpha = \frac{2\pi d}{\lambda} \cos\theta + \phi$.

$$\therefore H_T = NH \left[\frac{\sin N\left(\frac{\pi d}{\lambda} \cos\theta + \frac{\phi}{2}\right)}{N \sin\left(\frac{\pi d}{\lambda} \cos\theta + \frac{\phi}{2}\right)} \right] \quad \text{...........(5.17)}$$

5.13 Broadside Array

This is is an array in which the currents in the elements are in phase. This implies that $\phi = 0$.

$$\therefore H_T = NH \left[\frac{\sin N\left(\frac{\pi d}{\lambda} \cos\theta\right)}{N \sin\left(\frac{\pi d}{\lambda} \cos\theta\right)} \right] \quad \text{...........(5.18)}$$

$$= NHf(\theta),$$

where $f(\theta) = \dfrac{\sin N\left(\dfrac{\pi d}{\lambda}\cos\theta\right)}{N\sin\left(\dfrac{\pi d}{\lambda}\cos\theta\right)}$ ....(5.19)

When $\theta = 90°$ or $270°$,

$f(\theta) = 1$ and $H_T = NH$.

5.14 Polar Diagram of Broadside Array

Assume that the spacing between the antennas is $\dfrac{\lambda}{2}$. The polar diagram of the broadside array is given in Fig. 5.9.

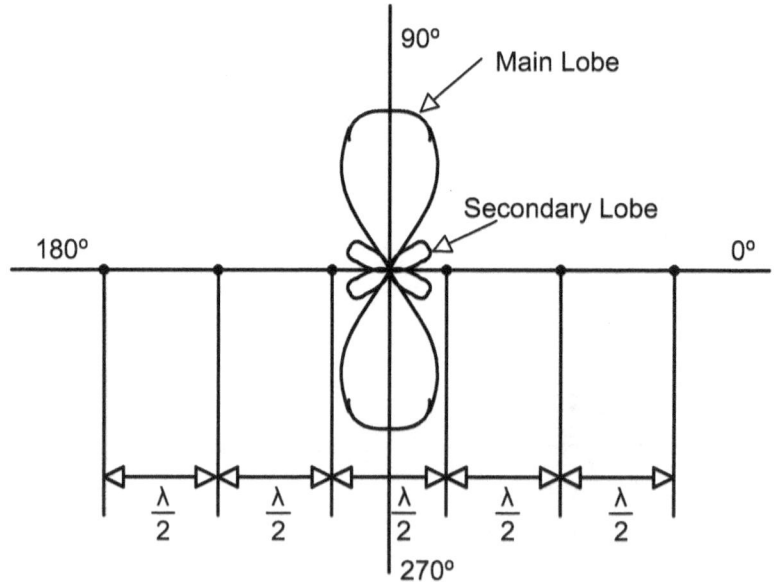

Fig. 5.9: Radiation pattern of broadside array

5.15 Beam Angle of Broadside Array

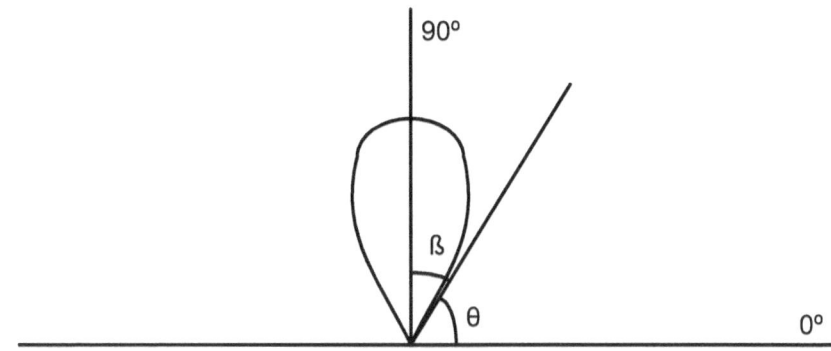

Fig. 5.10: Beam angle of broadside array

From Equation (5.19),

$$f(\theta) = \frac{\sin N\left(\dfrac{\pi d}{\lambda}\cos\theta\right)}{N\sin\left(\dfrac{\pi d}{\lambda}\cos\theta\right)}.$$

$\cos\theta = \sin\beta$.

$$\therefore f(\beta) = \frac{\sin N\left(\dfrac{\pi d}{\lambda}\sin\beta\right)}{N\sin\left(\dfrac{\pi d}{\lambda}\sin\beta\right)}.$$

The edge of the beam is where $f(\beta) = 0$.

$$\therefore \sin N\left(\frac{\pi d \sin\beta}{\lambda}\right) = 0$$

$$\therefore \frac{N\pi d \sin\beta}{\lambda} = \pi, 2\pi, 3\pi, \text{ etc.}$$

Consider the first case:

$$\frac{N\pi d \sin \beta_1}{\lambda} = \pi$$

$$\therefore \sin \beta_1 = \frac{\lambda}{Nd}.$$

$$\therefore \beta_1 = \sin^{-1}\left(\frac{\lambda}{Nd}\right).$$

If N is large, then $\sin \beta_1 \approx \beta_1$.

$$\therefore \beta_1 = \frac{\lambda}{Nd}$$

Beam angle $= 2\beta_1 = \dfrac{2\lambda}{Nd}$ ....(5.20)

If $d = \dfrac{\lambda}{2}$, then:

$$2\beta_1 = \frac{4\lambda}{N\lambda} = \frac{4}{N} \text{ radians.}$$

If $N = 2$, $2\beta_1 = 2 \text{ radians} = 115°$.

5.16 End-fire Array

In this array the phase angle of the current in successive elements is made equal to the phase lag due to the spacing of the elements. The implication of this is that in one direction all the fields arrive in phase. If this direction is $\theta = 0$, then we have:

$$\phi = -\frac{2\pi d}{\lambda}.$$

From Equation (5.17),

$$\therefore H_T = \frac{NH \sin N\left(\frac{\pi d}{\lambda}\cos\theta - \frac{\pi d}{\lambda}\right)}{N \sin\left(\frac{\pi d}{\lambda}\cos\theta - \frac{\pi d}{\lambda}\right)}$$

$$= \frac{NH \sin N\left(\frac{\pi d}{\lambda}\right)(\cos\theta - 1)}{N \sin\left(\frac{\pi d}{\lambda}\right)(\cos\theta - 1)} \quad\quad\quad\quad (5.21)$$

$$\therefore f(\theta') = \frac{\sin\left(\frac{N\pi d}{\lambda}\right)(1-\cos\theta')}{N \sin\left(\frac{\pi d}{\lambda}\right)(1-\cos\theta')} \quad\quad\quad\quad (5.22)$$

When $\theta' = 0$, $f(\theta') = 1$.

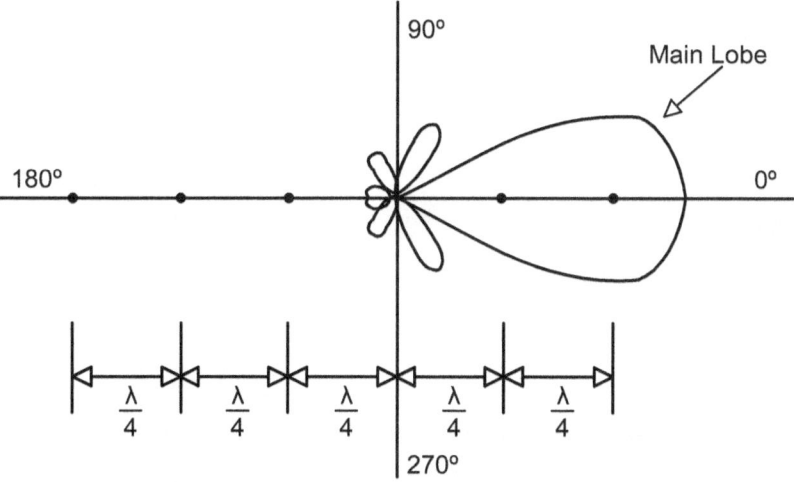

Fig. 5.11: Polar diagram for the end-fire array

5.17 Beam angle of the end-fire array

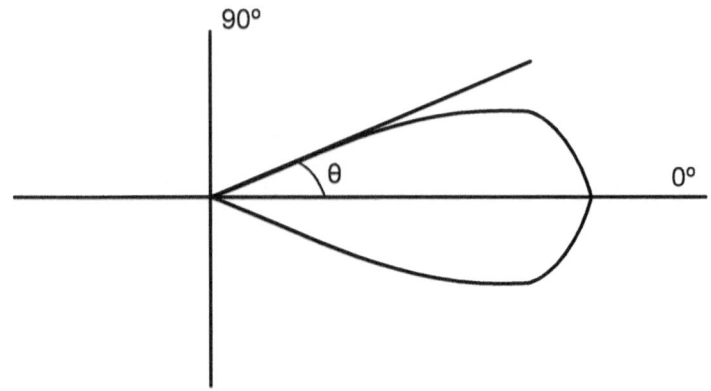

Fig. 5.12: Beam angle of the end-fire array

The zeros occur when $f(\theta') = 0$.

$$\therefore \frac{\sin\left(\frac{N\pi d}{\lambda}\right)(1-\cos\theta')}{N\sin\left(\frac{\pi d}{\lambda}\right)(1-\cos\theta')} = 0$$

$$\sin\left(\frac{N\pi d}{\lambda}\right)(1-\cos\theta') = 0$$

$$\therefore \left(\frac{N\pi d}{\lambda}\right)(1-\cos\theta') = \pi, 2\pi, 3\pi, \text{ etc.}$$

$$\therefore (1-\cos\theta') = \frac{\lambda}{Nd}, \frac{2\lambda}{Nd}, \frac{3\lambda}{Nd}, \ldots$$

$$\therefore \cos\theta' = \left(1-\frac{\lambda}{Nd}\right), \left(1-\frac{2\lambda}{Nd}\right), \left(1-\frac{3\lambda}{Nd}\right), \ldots$$

The first zero occurs when $\cos\theta'_1 = \left(1-\frac{\lambda}{Nd}\right)$.

The beam angle is given by:

Antennas

$$2\theta_1' = 2\cos^{-1}\left(1 - \frac{\lambda}{Nd}\right).$$

$$\cos^2\theta_1' = 1 - \sin^2\theta_1'$$

$$\therefore 1 - \sin^2\theta_1' = \left(1 - \frac{\lambda}{Nd}\right)^2$$

$$\sin^2\theta_1' = 1 - \left(1 - \frac{\lambda}{Nd}\right)^2$$

$$= 1 - \left(1 - \frac{2\lambda}{Nd} + \frac{\lambda^2}{N^2 d^2}\right)$$

$$= \frac{2\lambda}{Nd}\left(1 - \frac{\lambda}{2Nd}\right).$$

If N is large, then $\dfrac{\lambda}{2Nd} \ll 1$.

$$\therefore \sin^2\theta_1' = \frac{2\lambda}{Nd}.$$

Also if N is large, then θ_1' will be small.

So $\sin\theta_1' \approx \theta_1'$.

$$\therefore \theta_1'^2 = \frac{2\lambda}{Nd}.$$

Hence, $\theta_1' = \sqrt{\dfrac{2\lambda}{Nd}}$ radians.

Beam angle $= 2\theta_1' = 2\sqrt{\dfrac{2\lambda}{Nd}}$ ....(5.23)

If $d = \dfrac{\lambda}{2}$, then:

Beam angle $= 2\theta_1' = \dfrac{4}{\sqrt{N}}$ radians.

If $N = 2$, then $2\theta_1' = 2.828 \text{ radians} = 162°$.

5.18 Rhombic Antenna

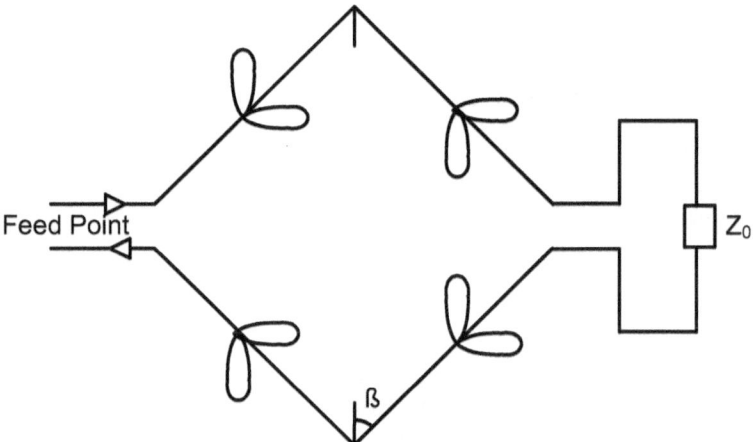

Fig. 5.13: Rhombic antenna

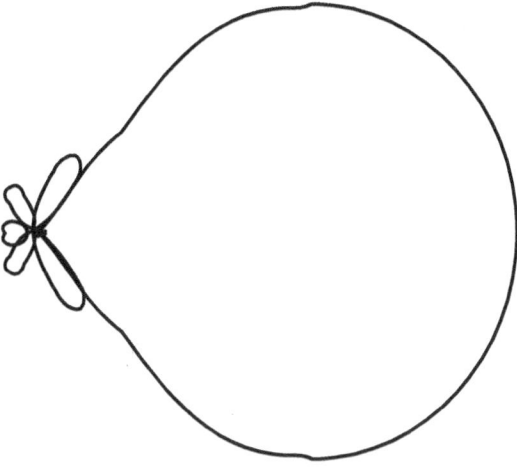

Fig. 5.14: Rhombic antenna radiation pattern

A rhombic antenna is a wide band directional antenna consisting of four non-resonant wires each several wavelengths long (2 to 8 λ) and arranged in the form of a rhombus. The lobe angle can be varied by adjusting the length of each radiator. The antenna has greater directivity than a single wire and can be terminated by an appropriate value of resistor to ensure non-resonance and a wide bandwidth. Since it is usually terminated in the characteristic impedance of the conductors, only forward waves are present and the efficiency is limited to about 50%. Rhombic antennas are used at high frequencies (3 to 30 MHz). The main advantage is that the input impedance and radiation pattern do not change rapidly over a considerable frequency range as compared with resonant dipoles. It is used in commercial point-to-point communication.

5.19 Yagi-Uda Antenna

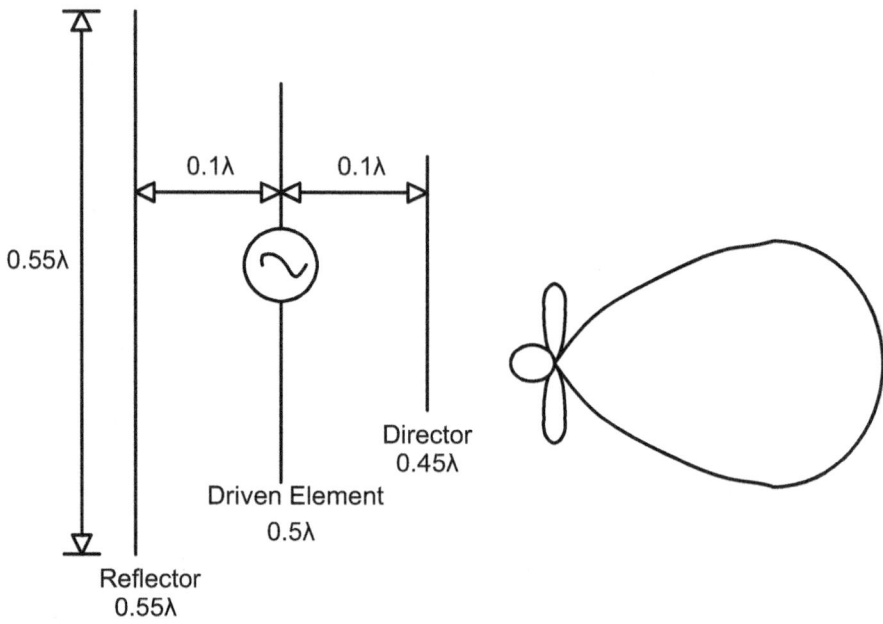

Fig. 5.15: Yagi-Uda antenna and radiation pattern

A Yagi-Uda antenna is a form of end-fire array consisting of one driven element and one or more parasitic elements excited by the field of the driven element. The driven element is usually a half-wave dipole. The single rear reflector is made slightly longer while the director elements are made slightly shorter than the driven element. The Yagi-Uda array produces one main lobe of radiation along the bisector of the element and forward from the reflector. It is used as a High Frequency (HF) transmitting antenna and a Very High Frequency (VHF) TV receiving antenna. As the radiators are brought closer to the driven element, the front-to-back ratio of the lobes can be improved. However, this has the adverse effect of lowering the input impedance. A separation of 0.1λ is the optimum value.

A Yagi-Uda antenna has the following advantages:
- It is compact.
- It is relatively broadband.
- It has good unidirectional radiation pattern.

5.20 Antennas with parabolic reflectors

The microwaves produced by Ultra-High Frequency (UHF) sources behave, in many respects, like light waves. It is possible to apply some of the principles of optics to UHF radiation. An example is the use of a parabolic reflector. Fig. 5.16(a) shows a parabola, CAD, whose focus is at F and whose axis is AB. The geometry of the parabola is such that FP + PP' = FQ + QQ' = FR + RR' = a constant.

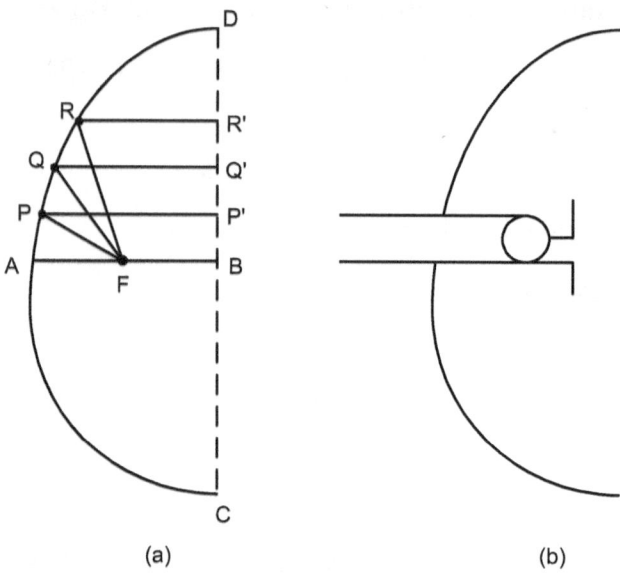

Fig. 5.16: a) Geometry of a parabola b) Antenna with parabolic reflector

If a source of radiation is placed at the focus, then all waves coming from the source and reflected by this parabola will travel the same distance by the time they reach the line DBC. Hence, all such waves will be in phase. This leads to a high degree of directivity.

The primary antenna is placed at the focus of the paraboloid for best results in transmission or reception. The direct radiation from the feed, which is not reflected by the paraboloid tends to spread out in all directions and hence partially spoils the directivity. A number of methods are used to prevent this. One of the methods is to use a small dipole array at the focus such as a Yagi-Uda or an end-fire array pointing at the parabolic reflector [1].

Antennas

5.21 Examples on Antennas

Example 1

An end-fire array of two elements is required to service an area which is 65 km wide at a radius of 30 km. The signal strength over this area is not to fall below 3dB of the maximum. Find the spacing between the elements in terms of the wavelength and hence, the phase difference of the current in the two elements.

Solution

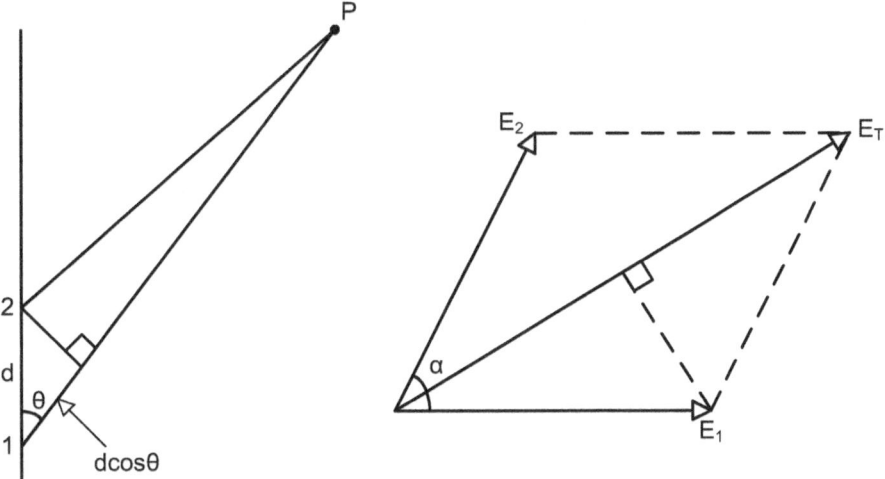

Fig. 5.17: Field due to two antennas in end-fire array

As shown in Fig. 5.17, the spacing between the two antenna elements is denoted by d.

Let E Volts/metre represent the field at point P due to each antenna.

The total field at P is given by:

$$E_T = 2E \cos\left(\frac{\alpha}{2}\right),$$

where α is the phase difference in the two fields.

α is made up of two components:

- Path difference of $d\cos\theta$ which leads to a phase angle difference of $\dfrac{2\pi d \cos\theta}{\lambda}$ radians.

- Phase angle difference in the two currents. For the end-fire array, this is equal to $-\dfrac{2\pi d}{\lambda}$.

Hence, $E_T = 2E \cos\left(\dfrac{\pi d \cos\theta}{\lambda} - \dfrac{\pi d}{\lambda}\right)$

$$= 2E \cos \frac{\pi d}{\lambda}(\cos\theta - 1)$$

$$= 2E \cos \frac{\pi d}{\lambda}(1 - \cos\theta) \quad\quad\quad\quad\quad\quad\quad\quad\quad\quad\quad\quad (5.24)$$

Antennas

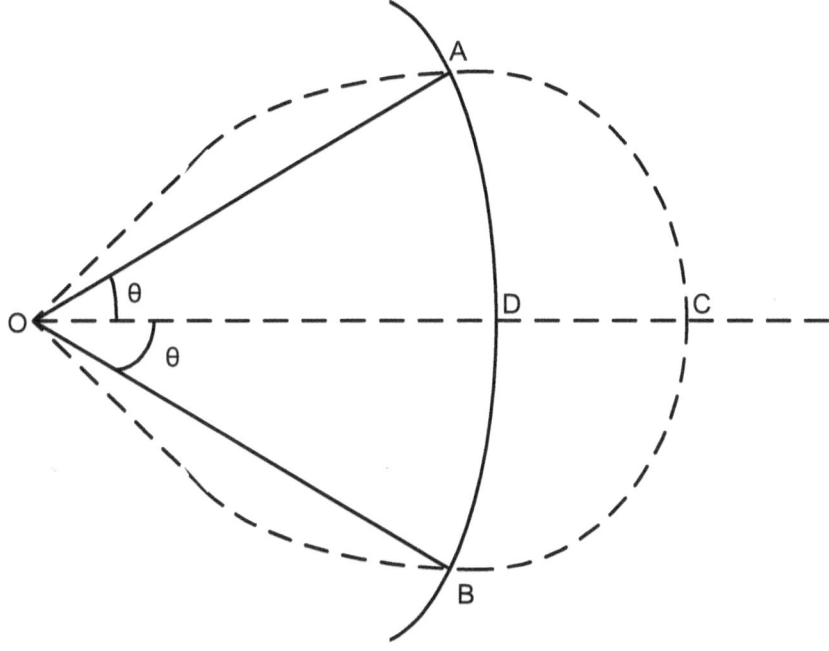

Fig. 5.18: Service area of the antennas in end-fire array

In Fig. 5.18, the arc ADB = 65km.

The radius of the service area is 30km.

Hence, the circumference of the service area will be $2\pi \times 30$ km.

$$\frac{2\theta}{360} = \frac{65}{2\pi \times 30}$$

$$\therefore \theta = \frac{65 \times 360}{2\pi \times 30 \times 2}$$

$$= 62.07°.$$

At points A and B, the field strengths are 3dB below the maximum value of $2E$.

$$\therefore 2E\cos\frac{\pi d}{\lambda}(1-\cos\theta) = \frac{2E}{\sqrt{2}}$$

$$2\cos\frac{\pi d}{\lambda}(1-\cos\theta) = 1.414$$

$$\cos\frac{\pi d}{\lambda}(1-\cos\theta) = 0.707$$

$$\frac{\pi d}{\lambda}(1-\cos\theta) = 45°$$

$$= 0.785 \text{ radians.}$$

But $\theta = 62.07°$.

$$\therefore \frac{\pi d}{\lambda}(1-\cos 62.07°) = 0.785$$

$$\frac{\pi d}{\lambda}(1-0.468) = 0.785$$

$$\frac{d}{\lambda} = 0.47$$

$$d = 0.47\lambda$$

Hence, the spacing between the elements is $d = 0.47\lambda$.

For an end-fire array, the phase angle difference in the two currents is $-\frac{2\pi d}{\lambda}$.

But $\frac{d}{\lambda} = 0.47$.

Hence, the phase angle difference is $-2\pi \times 0.47 = -0.94\pi$ radians $= -169.2°$.

Example 2

A broadside array of two elements is required to service an area which is 100km wide at a radius of 40km. The signal strength over this area is not to fall below 3dB of the maximum. Find the spacing between the elements in terms of the wavelength.

Solution

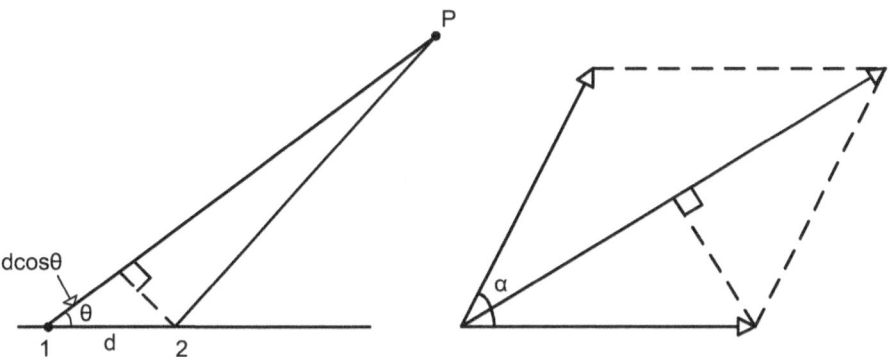

Fig. 5.19: Field due to two antennas in broadside array

As shown in Fig. 5.19, the spacing between the two antenna elements is denoted by d.

Let E Volts/metre represent the field at point P due to each antenna.
The total field at P is given by:

$$E_T = 2E \cos\left(\frac{\alpha}{2}\right),$$

where α is the phase difference in the two fields.
α is made up of two components:

- Path difference of $d\cos\theta$ which leads to a phase angle difference of $\dfrac{2\pi d \cos\theta}{\lambda}$ radians.

- Phase angle difference in the two currents. For the broadside array, this is equal to zero since the currents are in phase.

Hence, $E_T = 2E\cos\left(\dfrac{\pi d \cos\theta}{\lambda}\right)$

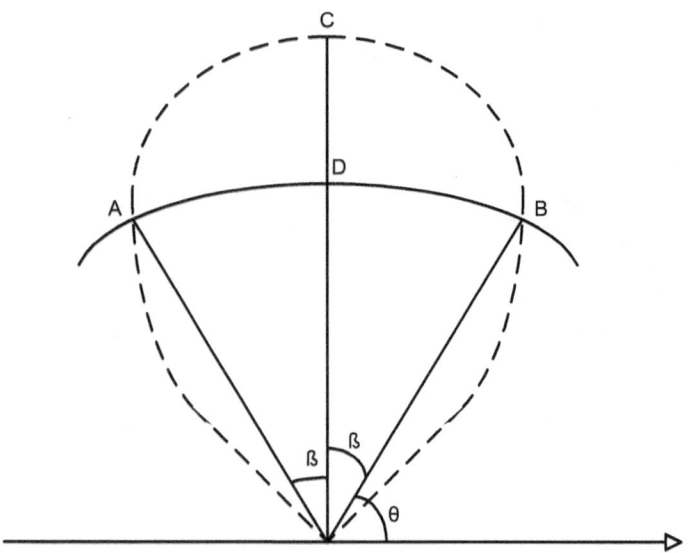

Fig. 5.20: Service area of the antennas in broadside array

In Fig. 5.20, the arc ADB = 100km.

The radius of the service area is 40km.

Hence, the circumference of the service area will be $2\pi \times 40$ km.

$$\dfrac{2\beta}{360} = \dfrac{100}{2\pi \times 40}$$

$$\therefore \beta = \frac{100 \times 360}{2\pi \times 40 \times 2}$$

$$= 71.6°.$$

$$\theta = 90° - \beta = 18.4°.$$

At points A and B, the field strengths are 3dB below the maximum value of $2E$.

$$\therefore 2E \cos\left(\frac{\pi d \cos\theta}{\lambda}\right) = \frac{2E}{\sqrt{2}}$$

$$2\cos\left(\frac{\pi d \cos\theta}{\lambda}\right) = 1.414$$

$$\cos\left(\frac{\pi d \cos\theta}{\lambda}\right) = 0.707$$

$$\frac{\pi d \cos\theta}{\lambda} = 45°$$

$$= 0.785 \text{ radians.}$$

But $\theta = 18.4°$.

$$\therefore \frac{\pi d}{\lambda}(\cos 18.4°) = 0.785$$

$$\frac{\pi d}{\lambda}(0.949) = 0.785$$

$$\frac{d}{\lambda} = 0.26$$

$$d = 0.26\lambda$$

Hence, the spacing between the elements is $d = 0.26\lambda$.

Example 3

An antenna array consists of 4 vertical elements spaced $\frac{\lambda}{4}$ apart in a straight line and energized by currents which are successively 90° out of phase. Derive from first principles, and sketch, the approximate shape of the polar diagram of the array.

Solution

The antenna array compirising of the four elements is illustrated in Fig. 5.21.

The path difference between two antennas is $d\cos\theta$. This leads to a phase difference of $\frac{2\pi d \cos\theta}{\lambda}$.

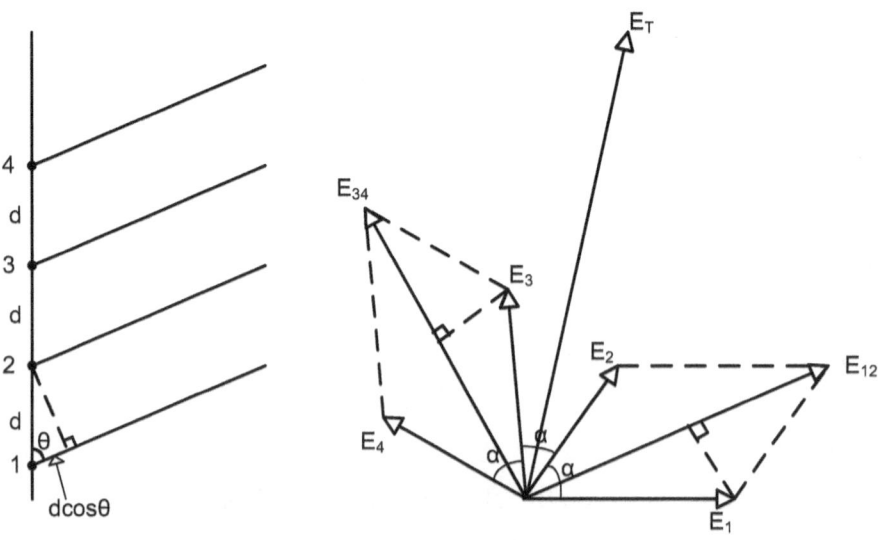

Fig. 5.21: Field due to array of 4 antennas

The phase difference in currents of successive antennas $= \dfrac{\pi}{2}$.

Hence, the total phase difference between the currents in successive antennas is given by:

$$\alpha = \dfrac{2\pi d \cos\theta}{\lambda} + \dfrac{\pi}{2}.$$

$$E_{12} = 2E \cos\left(\dfrac{\alpha}{2}\right),$$

where $E =$ field strength of each antenna.

Also, $E_{34} = 2E \cos\left(\dfrac{\alpha}{2}\right).$

Hence, total field strength is given by:

$$E_T = 2E_{12} \cos\alpha = 4E \cos\left(\dfrac{\alpha}{2}\right) \cos\alpha \quad\text{...............................(5.25)}$$

But $\alpha = \dfrac{2\pi d \cos\theta}{\lambda} + \dfrac{\pi}{2}.$

$$\therefore E_T = 4E \cos\left(\dfrac{\pi d \cos\theta}{\lambda} + \dfrac{\pi}{4}\right) \cos\left(\dfrac{2\pi d \cos\theta}{\lambda} + \dfrac{\pi}{2}\right).$$

But $d = \dfrac{\lambda}{4}.$

Hence, $\alpha = \dfrac{\pi \cos\theta}{2} + \dfrac{\pi}{2} = \dfrac{\pi}{2}(\cos\theta + 1).$ ....(5.26)

$$\therefore E_T = 4E \cos\left(\dfrac{\pi \cos\theta}{4} + \dfrac{\pi}{4}\right) \cos\left(\dfrac{\pi \cos\theta}{2} + \dfrac{\pi}{2}\right)$$

$$= 4E \cos\left[\dfrac{\pi}{4}(\cos\theta + 1)\right] \cos\left[\dfrac{\pi}{2}(\cos\theta + 1)\right]$$

From Equation (5.25), when $\alpha = 0$, $E_T = 4E$.

From Equation (5.26), when $\alpha = 0$, $\cos\theta = -1$, $\therefore \theta = 180°$.

Also from Equation (5.26), when $\theta = 0°$, $\alpha = \pi$.

From Equation (5.25), when $\alpha = \pi$, $E_T = 0$.

When $\theta = 45°$, $\alpha = \dfrac{\pi}{2}(\cos 45° + 1) = 153.6°$,

$E_T = 4E\cos(76.8°)\cos 153.6° = -0.82E$.

When $\theta = 90°$, $\alpha = \dfrac{\pi}{2}(\cos 90° + 1) = 90°$, $E_T = 4E\cos(45°)\cos 90° = 0$.

When $\theta = 135°$, $\alpha = \dfrac{\pi}{2}(\cos 135° + 1) = 26.4°$,

$E_T = 4E\cos(13.2°)\cos 26.4° = 3.5E$.

When $\theta = 225°$, $\alpha = \dfrac{\pi}{2}(\cos 135° + 1) = 26.4°$,

$E_T = 4E\cos(13.2°)\cos 26.4° = 3.5E$.

When $\theta = 270°$, $\alpha = \dfrac{\pi}{2}(\cos 270° + 1) = 90°$,

$E_T = 4E\cos(45°)\cos 90° = 0$.

From Equation (5.25), $E_T = 2E_{12}\cos\alpha = 4E\cos\left(\dfrac{\alpha}{2}\right)\cos\alpha$.

$\therefore \dfrac{dE_T}{d\alpha} = 4E\left[-\dfrac{1}{2}\sin\left(\dfrac{\alpha}{2}\right)\cos\alpha - \cos\left(\dfrac{\alpha}{2}\right)\sin\alpha\right]$.

For maximum or minimum,

$\dfrac{dE_T}{d\alpha} = 0$.

$\therefore 4E\left[-\dfrac{1}{2}\sin\left(\dfrac{\alpha}{2}\right)\cos\alpha - \cos\left(\dfrac{\alpha}{2}\right)\sin\alpha\right] = 0$

$\dfrac{1}{2}\sin\left(\dfrac{\alpha}{2}\right)\cos\alpha = -\cos\left(\dfrac{\alpha}{2}\right)\sin\alpha$

$\dfrac{1}{2}\tan\left(\dfrac{\alpha}{2}\right) = -\tan\alpha$

$= \dfrac{-2\tan\left(\dfrac{\alpha}{2}\right)}{1-\tan^2\left(\dfrac{\alpha}{2}\right)}$

$\therefore \dfrac{1}{2}\tan\left(\dfrac{\alpha}{2}\right) - \dfrac{1}{2}\tan^3\left(\dfrac{\alpha}{2}\right) = -2\tan\left(\dfrac{\alpha}{2}\right)$

$\tan^3\left(\dfrac{\alpha}{2}\right) - 5\tan\left(\dfrac{\alpha}{2}\right) = 0$

$\tan\left(\dfrac{\alpha}{2}\right)\left(\tan^2\left(\dfrac{\alpha}{2}\right) - 5\right) = 0$

$\therefore \tan\left(\dfrac{\alpha}{2}\right) = 0 \text{ or } \tan\left(\dfrac{\alpha}{2}\right) = \pm\sqrt{5}$

$\tan\left(\dfrac{\alpha}{2}\right) = 0 \text{ or } \tan\left(\dfrac{\alpha}{2}\right) = \pm 2.236$

When $\tan\left(\dfrac{\alpha}{2}\right) = \pm 2.236$,

$\dfrac{\alpha}{2} = 65.9°$

$\alpha = 131.8°$

But $\alpha = \dfrac{\pi \cos\theta}{2} + \dfrac{\pi}{2}$.

$\theta = \cos^{-1}\left(\dfrac{2\alpha}{\pi} - 1\right)$

$\theta = \cos^{-1}(0.4645)$

$\theta = 62.3°$.

$E_T = 4E \cos\left(\dfrac{\alpha}{2}\right) \cos\alpha$

When $\alpha = 131.8°$,

$E_T = 4E \cos 65.9° \cos 131.8°$

$ = 1.08E$

Fig. 5.22 shows the shape of the polar diagram for the array.

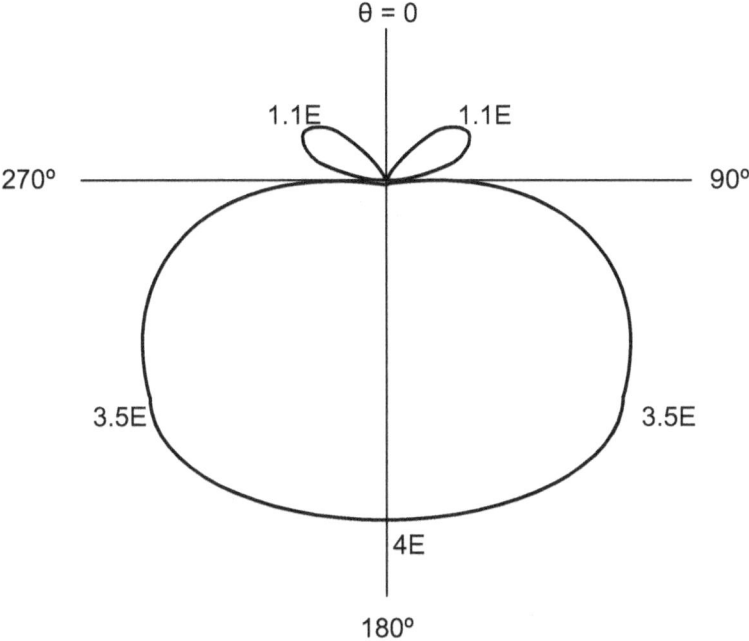

Fig. 5.22: Polar diagram of an array of 4 antennas

Example 4

An elementary doublet is 10 cm long. If a current of 2 amps at a frequency of 10 MHz is flowing through it, calculate the field strength 10 km away from the doublet in a direction of maximum radiation.

Solution

The average power radiated by a doublet is given in Equation (5.4) as:

$$P_a = 120\pi \left(\frac{I \delta l \sin \theta}{2r\lambda} \right)^2.$$

$$\therefore EH = 120\pi \left(\frac{I\delta l \sin\theta}{2r\lambda}\right)^2,$$

where E = electric component of the radiation field, and
H = magnetic component of the radiation field.

But $H = \dfrac{E}{120\pi}$.

$$\therefore E^2 = (120\pi)^2 \left(\frac{I\delta l \sin\theta}{2r\lambda}\right)^2.$$

$$E = 120\pi \left(\frac{I\delta l \sin\theta}{2r\lambda}\right).$$

$I = 10$ A.

$\delta l = 10$ cm $= 0.1$ m

$r = 10$ km $= 10^4$ m

$f = 10$ MHz $= 10^7$ Hz

$$\lambda = \frac{c}{f} = \frac{3\times 10^8}{10^7}$$

$= 30$ m

$$\therefore E = 120\pi \left(\frac{10 \times 0.1 \sin\theta}{2 \times 10^4 \times 30}\right)$$

$= 6.3 \times 10^{-4} \sin\theta$ V/m.

In the direction of maximum radiation, $\theta = 0°$.

$\therefore E_{max} = 6.3 \times 10^{-4}$ V/m.

Example 5

Find the field strength in the directions $0°$, $90°$, $180°$, and $270°$ for a two element end fire array in which the spacing is a quarter of a wavelength. Plot the polar diagram.

Solution

From Equation (5.24), the total field strength for a two element end-fire array is given by:

$$E_T = 2E \cos \frac{\pi d}{\lambda}(1 - \cos\theta).$$

But $d = \frac{\lambda}{4}$.

$$\therefore E_T = 2E \cos \frac{\pi}{4}(1 - \cos\theta).$$

When $\theta = 0°$, $E_T = 2E \cos \frac{\pi}{4}(1 - \cos 0°) = 2E \cos 0 = 2E$.

When $\theta = 90°$, $E_T = 2E \cos \frac{\pi}{4}(1 - \cos 90°) = 2E \cos \frac{\pi}{4} = 1.41E$.

When $\theta = 180°$, $E_T = 2E \cos \frac{\pi}{4}(1 - \cos 180°) = 2E \cos \frac{\pi}{2} = 0$.

When $\theta = 270°$, $E_T = 2E \cos \frac{\pi}{4}(1 - \cos 270°) = 2E \cos \frac{\pi}{4} = 1.41E$.

When $\theta = 45°$, $E_T = 2E \cos \frac{\pi}{4}(1 - \cos 45°) = 2E \cos \frac{0.29\pi}{4} = 1.95E$.

When $\theta = 135°$, $E_T = 2E \cos \frac{\pi}{4}(1 - \cos 135°) = 2E \cos \frac{1.71\pi}{4} = 0.46E$.

When $\theta = 225°$, $E_T = 2E\cos\dfrac{\pi}{4}(1-\cos 225°) = 2E\cos\dfrac{1.71\pi}{4} = 0.46E$.

When $\theta = 315°$, $E_T = 2E\cos\dfrac{\pi}{4}(1-\cos 315°) = 2E\cos\dfrac{0.29\pi}{4} = 1.95E$.

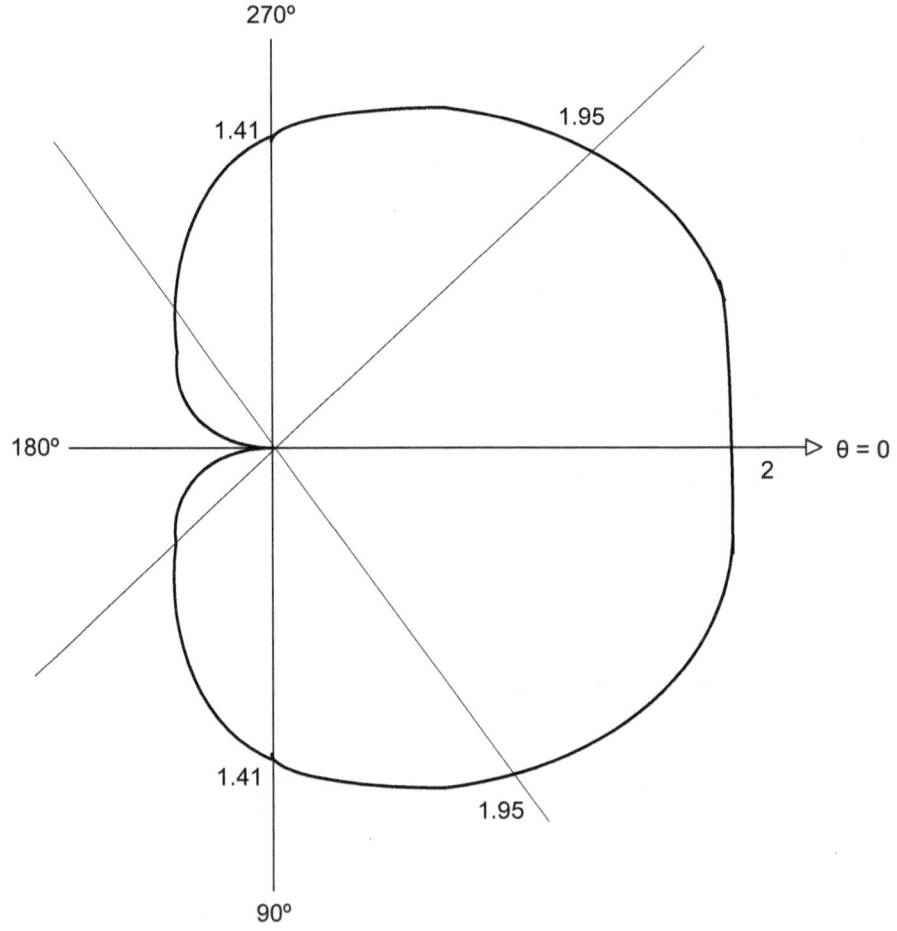

Fig. 5.23: Polar diagram of two element end-fire array

5.22 Questions on Antennas

Question 1

An end-fire array of 4 elements is required to service an area which is 50 km wide at a radius of 25 km. The signal strength over this area is not to fall below 3dB of the maximum. Find the spacing between the elements in terms of the wavelength and hence, the phase difference of the current in the successive elements.

Answer: 0.248λ ; $-89.3°$

Question 2

An broadside array of 4 elements is required to service an area which is 120 km wide at a radius of 50 km. The signal strength over this area is not to fall below 3dB of the maximum. Find the spacing between the elements in terms of the wavelength.

Answer: 0.316λ

Question 3

A receiving antenna at a height h_2 is placed at a distance d from a transmitting antenna at a height h_1. Show that the path difference, x, between a direct wave and a reflected wave at the receiver is given by:

$$x = \frac{2h_1 h_2}{d}.$$

Assume that the angle of reflection is small and the curvature of the earth can be neglected. Calculate the phase difference between the two waves if $h_1 = 200\,\text{m}$, $h_2 = 25\,\text{m}$, $d = 100\,\text{km}$, and the frequency is 1 MHz.

Answer: 0.12°

6

Unguided Media (Wireless)

6.1 Introduction

Unguided media or wireless communications transport electromagnetic waves without using a physical conductor. Instead signals are broadcast through air (or in a few cases water) and are available to anyone with a device capable of receiving them.

6.2 Frequency Allocation

The entire electromagnetic spectrum is split into different frequency groups with different uses or applications as illustrated in Fig. 6.1.

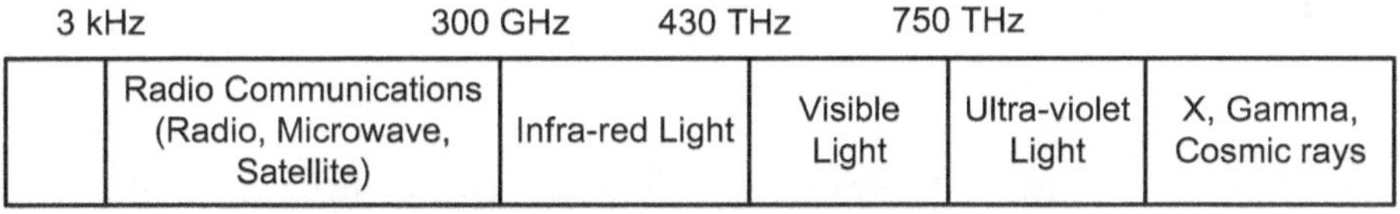

	3 kHz	300 GHz	430 THz	750 THz	
	Radio Communications (Radio, Microwave, Satellite)	Infra-red Light	Visible Light	Ultra-violet Light	X, Gamma, Cosmic rays

Fig. 6.1: The electromagnetic spectrum. [4]

The section of the electromagnetic spectrum defined as radio communication is divided into eight ranges called bands as shown in Fig.

6.2. Each band is regulated by Government Authorities. These bands are as follows:

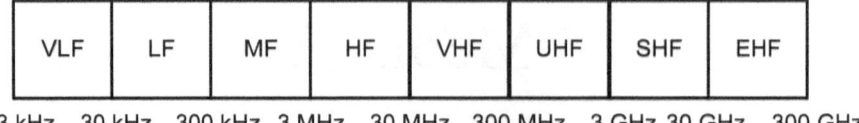

Fig. 6.2: The radio communications spectrum. [4]

- Very Low Frequency (VLF): 3 kHz – 30 kHz.
- Low Frequency (LF): 30 kHz – 300 kHz.
- Medium Frequency (MF): 300 kHz – 3 MHz.
- High Frequency (HF): 3 MHz – 30 MHz.
- Very High Frequency (VHF): 30 MHz – 300 MHz.
- Ultra High Frequency (UHF): 300 MHz – 3 GHz.
- Super High Frequency (SHF): 3 GHz – 30 GHz.
- Extremely High Frequency (EHF): 30 GHz – 300 GHz.

6.3 Types of Propagation

Radio wave transmission uses five different types of propagation, namely: [4]

- Surface Propagation
- Tropospheric Propagation
- Ionospheric Propagation
- Line-of-sight Propagation

- Space Propagation

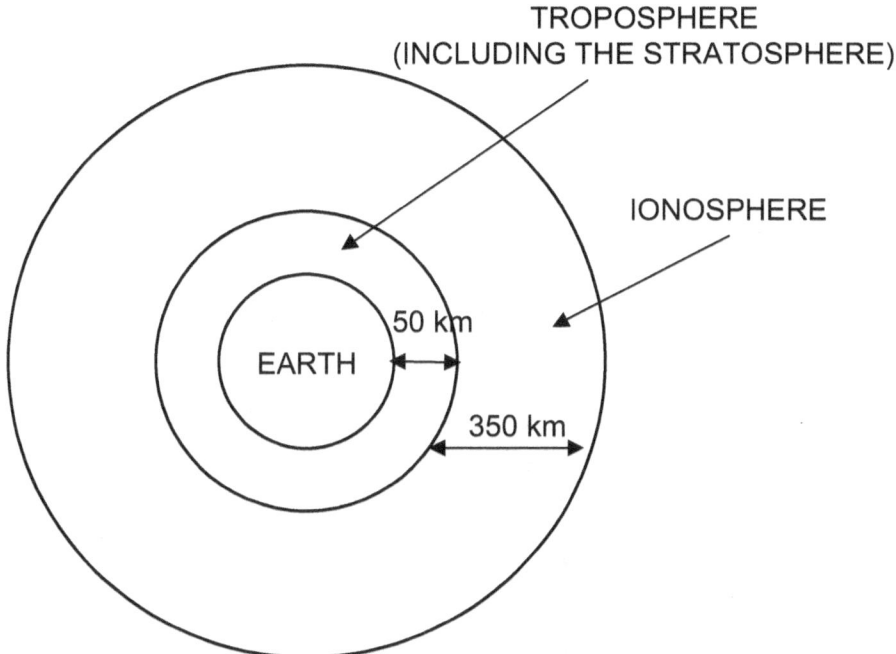

Fig. 6.3: The Earth with surrounding layers of atmosphere

The earth may be considered as being surrounded by two main layers of atmosphere – the Troposphere which includes the Stratosphere, and the Ionosphere as shown in Fig. 6.3.

The Troposphere is the portion of the atmosphere extending outward approximately 50 Km from the surface of the earth. It contains what is generally thought of as air. Clouds, wind and temperature variations and weather in general occur in the Troposphere. Also jet plane travel takes place in the Troposphere. The Ionosphere is the layer of the atmosphere above the Troposphere. It starts at about 50 Km from the surface of the earth and extends to about 400 Km from the surface of the earth. It contains electrically charged particles.

6.3.1 Surface Propagation

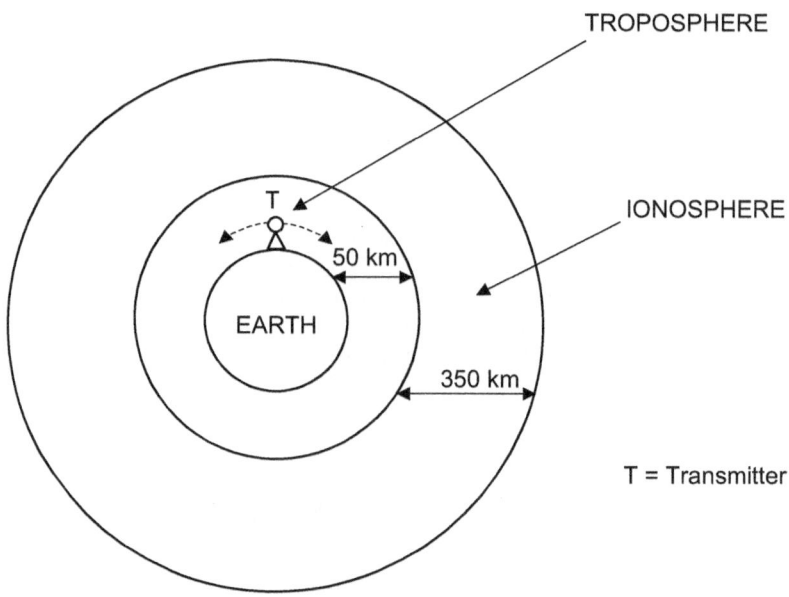

Fig. 6.4: Surface Propagation

The radio waves travel through the lowest portion of the atmosphere, hugging the earth. The radio signal comes from the transmitting antenna and follows the curvature of the earth. The distance covered depends on the amount of power transmitted and the terrain. Surface propagation can also take place in sea water. It uses Very Low Frequency and Low Frequency signals in the bands of 3 KHz – 300 KHz.

6.3.2 Tropospheric Propagation

The signal is directed at an angle into the upper layers of the Troposphere where it is reflected back to the earth's surface. This allows for greater distance to be covered. If the transmitter T and receiver R are within line-of-sight then Line-of-Sight propagation can also apply. It uses Medium Frequency band (300 KHz – 3 MHz).

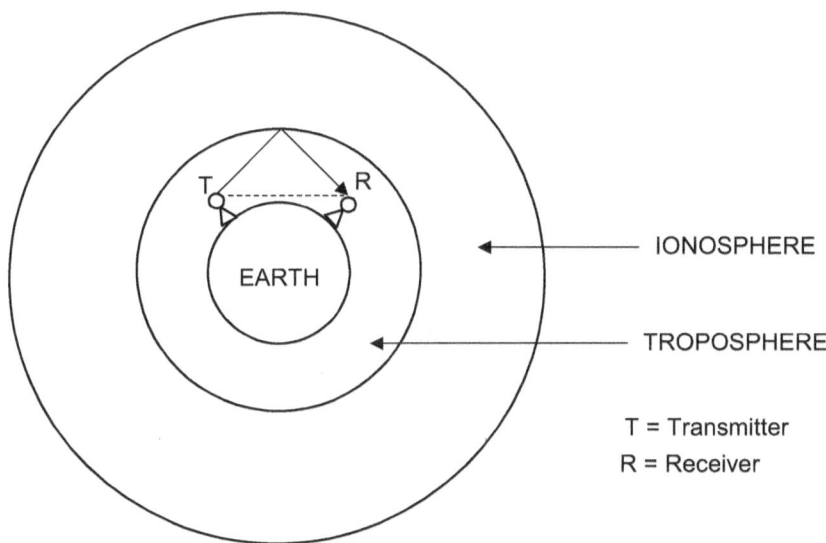

Fig. 6.5: Tropospheric Propagation

6.3.3 Ionospheric Propagation

High frequency radio waves are directed upwards into the atmosphere where they are reflected back to the earth. The density difference between the troposphere and the ionosphere causes the radio waves to change direction bending them back to the earth. It requires a high density of electrons in the ionization region. It allows for greater distances to be covered with lower power from the transmitter. It uses High Frequency band (3 MHz – 30 MHz).

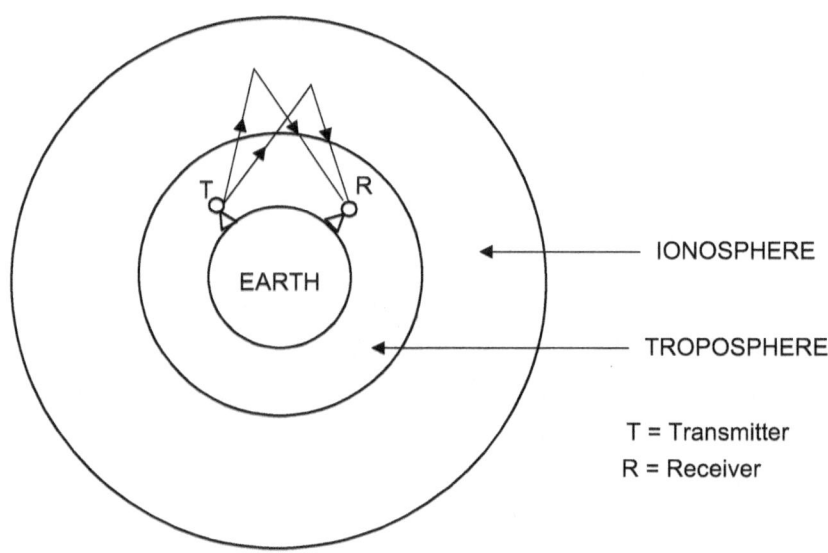

Fig. 6.6: Ionospheric Propagation

6.3.4 Line-of-sight Propagation

VHF and UHF signals are transmitted in straight line directly from antenna to antenna. The antennas must be directional facing each other. They must either be tall enough or close enough together not to be affected by the curvature of the earth. One difficulty encountered is that in addition to the direct waves there will be waves that are reflected from off the surface of the earth or from parts of the atmosphere. Such reflected waves that arrive at the receiving antenna later than the direct portion of the transmission can corrupt the received signal.

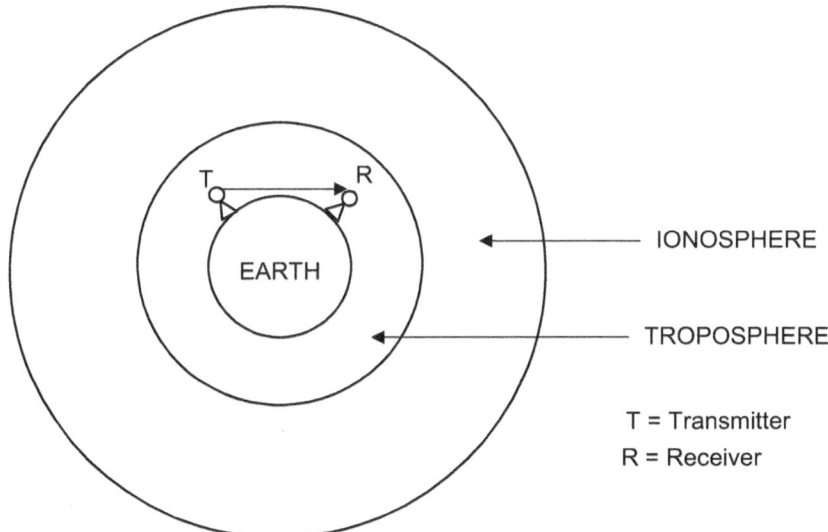

Fig. 6.7: Line-of-sight Propagation

6.3.5 Space Propagation (Satellite Communication)

Space propagation uses satellite relays in place of atmospheric reflections. In 1945, Arthur C. Clarke suggested that the principle of reflection of radio waves could be extended by placing a man-made reflector out in space in an orbit synchronized to the Earth's rotation (Geosynchronous or Geostationary Orbit GEO).

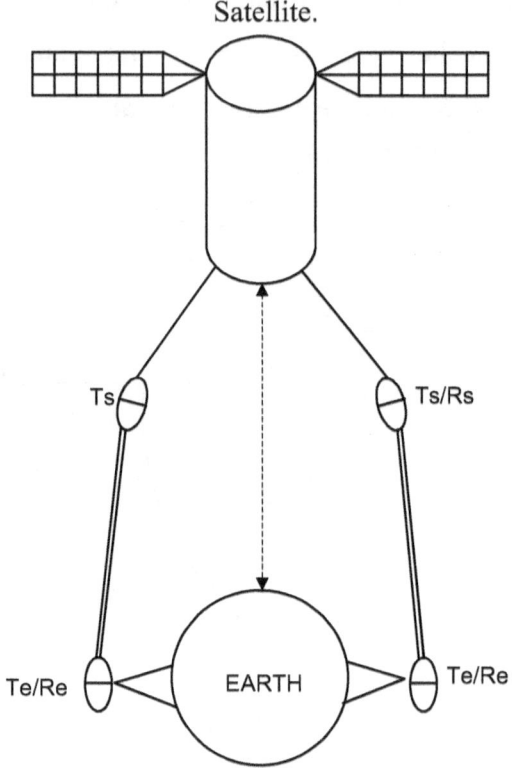

Fig. 6.8: Satellite Communication. [4]

A communication satellite is essentially a microwave repeater. It receives the energy beamed up at it by an earth station. It amplifies and returns it to earth at a frequency about 2 GHz away. The frequency shift prevents interference between the uplink and the downlink. The influence imposed on distance by the curvature of the earth is reduced. It can provide transmission capability to and from any location on earth, no matter how remote. This advantage makes high quality communication available to under-developed parts of the world without requiring a huge investment in ground based infrastructure.

6.4 Kepler's Three Laws of Planetary Motion

6.4.1 The law of ellipses

The first law states that the paths of the planets above the sun are elliptical in shape with the centre of the sun being located at one focus. This is referred to as the law of Ellipses.

6.4.2 The law of equal areas

The second law states that an imaginary line drawn from the centre of the sun to the centre of the planet will sweep out equal areas in equal intervals of time. This is the law of equal areas. This implies that the speed of motion of any planet will be constantly changing. A planet moves fastest when it is closest to the sun and slowest when it is furthest from the sun.

6.4.3 The law of harmonies

The third law states that the ratio of the squares of the periods is equal to the ratio of the cubes of their average distances from the sun. This is the law of Harmonies.

$$\frac{T^2}{R^3} = \text{constant}, \dots\dots\dots\dots\dots\dots(6.1)$$

where T = period of orbit, and

R = mean radius of orbit.

For earth, the value of the constant is 2.977×10^{-19} s^2/m^3.

For Mars, it is 2.975×10^{-19} s^2/m^3.

6.4.4 A demonstration of the third law

Consider a planet of mass M_p that orbits in a near circular motion about the sun of mass M_s.

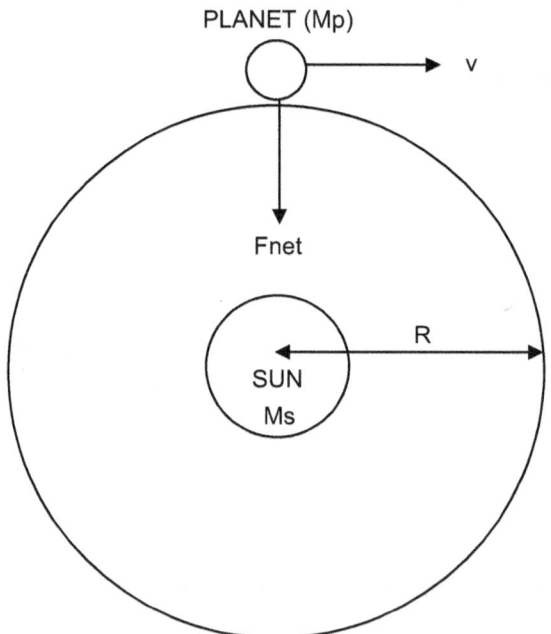

Fig. 6.9: A planet in near circular orbit around the sun

The net centripetal force acting on the planet is given by:

$$F_{net} = \frac{M_p v^2}{R}, \quad\quad\quad\quad\quad\quad\quad\quad\quad\quad\quad\quad\quad\quad\quad\quad\quad\quad\quad\quad (6.2)$$

where R = radius of the orbit, and

v = speed of the planet.

This net centripetal force is the result of the gravitational force which attracts the planet towards the sun and can be represented by:

$$F_{grav} = \frac{GM_p M_s}{R^2}, \quad\quad\quad\quad\quad\quad\quad\quad\quad\quad\quad\quad\quad\quad\quad\quad\quad\quad (6.3)$$

where G = universal gravitational constant = 6.67×10^{-11} Nm²/kg².

But $F_{net} = F_{grav}$.

$$\therefore \frac{M_p v^2}{R} = \frac{GM_p M_s}{R^2}.$$

$$\frac{v^2}{R} = \frac{GM_s}{R^2} \quad\quad\quad\quad\quad\quad\quad\quad\quad\quad\quad\quad\quad\quad\quad\quad\quad\quad\quad\quad (6.4)$$

Hence, acceleration $= \dfrac{GM_s}{R^2}$.

$$v^2 = \frac{GM_s}{R}. \quad\quad\quad\quad\quad\quad\quad\quad\quad\quad\quad\quad\quad\quad\quad\quad\quad\quad\quad\quad (6.5)$$

$$v = \sqrt{\frac{GM_s}{R}}. \quad\quad\quad\quad\quad\quad\quad\quad\quad\quad\quad\quad\quad\quad\quad\quad\quad\quad\quad (6.6)$$

The velocity of an object in a near circular orbit is given by:

$$v = \frac{2\pi R}{T}, \quad\quad\quad\quad\quad\quad\quad\quad\quad\quad\quad\quad\quad\quad\quad\quad\quad\quad\quad\quad\quad (6.7)$$

where T = period of orbit.

$$\therefore v^2 = \frac{4\pi^2 R^2}{T^2} . \quad \text{...(6.8)}$$

From Equations (6.5) and (6.8), we have:

$$\frac{GM_s}{R} = \frac{4\pi^2 R^2}{T^2} .$$

$$\therefore \frac{T^2}{R^3} = \frac{4\pi^2}{GM_s} . \quad \text{...(6.9)}$$

Since $G = 6.67 \times 10^{-11}$ Nm²/kg² and $M_s = 1.99 \times 10^{30}$ kg,

$$\frac{T^2}{R^3} = 2.974 \times 10^{-19} \text{ s}^2/\text{m}^3 . \quad \text{...(6.10)}$$

Equation (6.10) demonstrates Kepler's law of Harmonies. It applies to every planet regardless of the mass of the planet.

6.5 Fundamentals of satellite communication

Fig. 6.10 shows a typical satellite communication system which consists of the following: [6]
- A satellite in space.
- Several transmissions from transmitter ground earth stations to the satellite – UPLINK.
- Several transmissions from satellite to several receiver ground earth stations – DOWNLINK. The simplest earth stations are

receive-only terminals permanently interfaced to a single data user such as Domestic Satellite TV (DSTV) receivers.
- Gateway Earth stations which are capable of both transmission and reception. They have links to multiple data sources and users via one or more terrestrial networks.
- Satellite control network to maintain control operability by tracking and controlling the satellite.

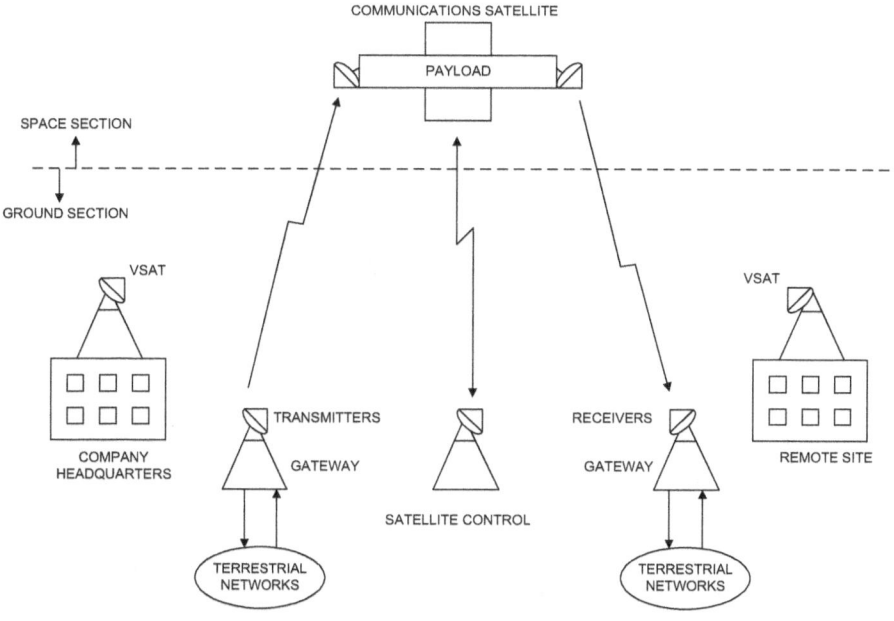

Fig. 6.10: A satellite communication system

6.5.1 Advantages of satellite communications. [6]

Satellite communications offer the following advantages:

- It offers global coverage irrespective of location or terrain.
- It offers a wide bandwidth and low error communications.
- It offers distance independent cost. For conventional terrestrial communication system, roll-out costs can be relatively low but generally increase proportional to the transmission distance. The roll-out costs of satellite communications may be significantly higher. However, the cost of transmitting to anywhere within the satellite coverage area is approximately constant.

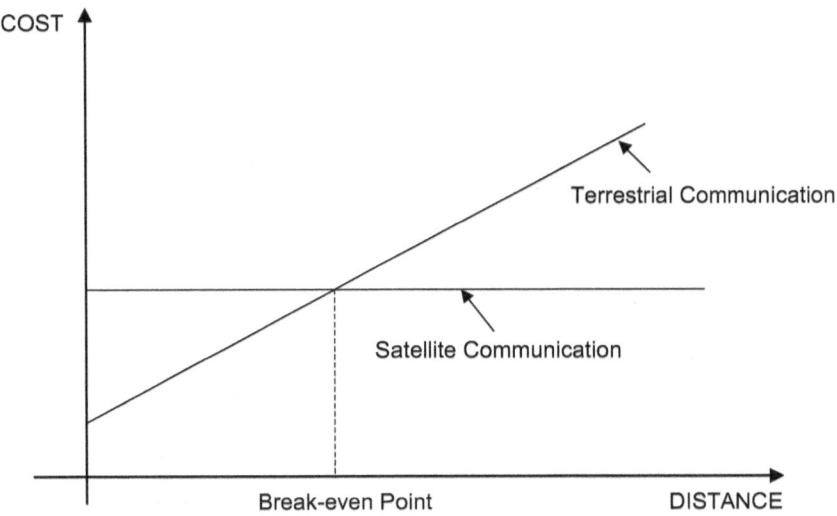

Fig. 6.11: Comparison of satellite and terrestrial communication costs

- It offers rapid installation of ground networks. Once the satellite is operational individual Earth stations can be activated quickly in response to demand for service.
    - It offers uniform service characteristics.

6.5.2 Disadvantages of satellite communications. [6]

Satellite communications have the following disadvantages:
- The cost of the system is high. The cost of purchasing a GEO Satellite can vary between $(100 – 500) million while the cost of launching the satellite can range from $(100 – 200) million.
- It has limited life time and maintenance difficulty. The life time of a satellite is about 10 years.
- The transit time delay associated with the round trip from Earth to Satellite to Earth is significant and generally requires echo suppression / cancellation.

6.6 Satellite orbits

The main types of satellite orbits are geostationary orbits (GEOs), low earth orbits (LEOs), and medium earth orbits (MEOs).

6.6.1 Geostationary orbit (GEO)

This is the most common orbit. It is a circular orbit about the equator at an altitude of about 37,786 km. The satellite has the same angular velocity as the earth with an orbital period of 24 hours. Hence the satellite is synchronized with the rotation of the Earth and appears stationary in the sky.

Advantages of GEO

- It offers 24 hours availability.
- It is fixed in the sky therefore no tracking is required except for very large antennas.
- It has no relative motion hence there is no Doppler shift.
- One satellite gives 42% Earth coverage hence three satellites give global coverage.

Disadvantages of GEO

- It has large altitude, hence long return path.
- It has a high power loss, typically about 200dB.
- It has a large propagation delay, typically between 240 and 280 ms. Hence it requires echo suppression.

6.6.2 Low/ Medium earth orbit (LEO/ MEO)

These are generally circular orbits with variable inclination. LEOs have altitudes of between 500 and 1,500 km. MEOs have altitudes of 8,000 – 12,000 km.

Advantages of LEO/MEO

- Low altitude hence low loss and low delay.

- Low power requirements for the satellite hence satellite can be smaller and lighter. This leads to cost savings.

Disadvantages of LEO/MEO

- Low altitude, hence smaller coverage per satellite. Thus more satellites are required to provide global coverage. For MEOs, 12 satellites are required to provide global coverage while LEOs require 80 satellites for complete coverage.
- Shorter period, hence low availability.
- Non- synchronous orbit, hence tracking is necessary and Doppler shift must be considered.
- The total cost for a constellation is still comparable to GEO global system.

6.7 Historical overview of satellite communication. [6]

- 1945 – Arthur C. Clark suggested that the principle of reflection of radio waves could be extended by placing a manmade reflector (Satellite) in an orbit synchronized to the earth's rotation.
- 1957 – Russia launched the first satellite with radio transmitter – SPUTNIK 1 in a low earth orbit.

- 1960 – USA launched two experimental satellites – ECHO 1 and ECHO 2.
- 1962 – USA launched two satellites TELSTAR 1 and RELAY 1. They relayed TV pictures.
- 1965 – The International Telecommunication Satellite Organization (INTELSAT) was formed with eleven member nations.
- 1965 – USA launched the first GEO satellite INTELSAT 1.
- 1967 – USA launched the second GEO satellite INTELSAT 2.
- 1969 – USA launched the third GEO satellite INTELSAT 3.
- 1970 – The three INTELSAT satellites were spaced 120 degrees apart to provide global communication, thus realizing Clark's prophecy.
- 1972 – Canada launched a domestic communication satellite using GEO orbit.
- 1975 – INTELSAT 4 was launched with solar cells to provide power.
- 1975 – India launched a communication satellite Aryabhata.
- 1976 – Indonesia launched a communication satellite.
- 1980 – INTELSAT 5 was launched.
- 1989 – INTELSAT 6 was launched.
- 1993 – INTELSAT 7 was launched.
- 2000 – Satellite uses Global Positioning System (GPS) for navigation.

- 2003 – Nigeria SAT 1 was launched for Nigeria by Russia.
- 2007 – NIGCOMSAT 1 was launched for Nigeria by China. It is used for broadcast, phone and internet services.

6.8 Examples on Satellite Communications

Example 1

A satellite orbits the earth at a height of 200 km above the surface of the earth. Determine the speed and orbital period of the satellite given the following:

Mass of the earth = 5.98×10^{24} kg,

Radius of the earth = 6.37×10^{6} m, and

Universal gravitation constant = 6.67×10^{-11} Nm²/kg².

Solution

The radius of the orbit is calculated as:

$R = h + r$

$\quad = 2 \times 10^{5} + 6.37 \times 10^{6}$

$\quad = 6.57 \times 10^{6}$ m.

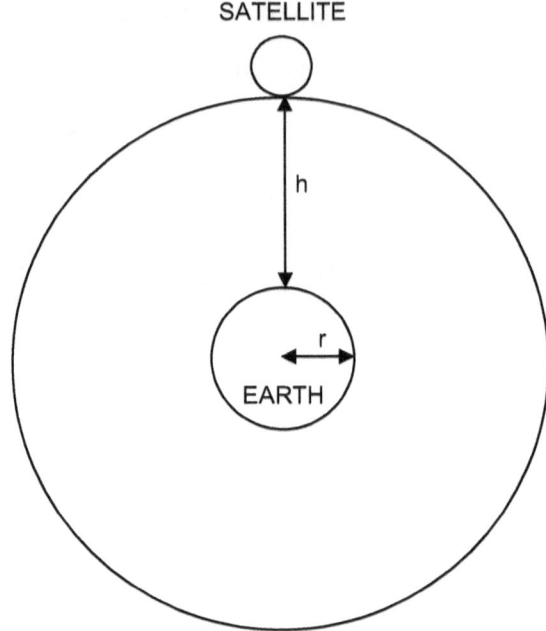

Fig. 6.12: Figure for Example 1

The velocity of the satellite is given by:

$$v = \sqrt{\frac{GM_e}{R}} \text{ m/s,}$$

where G = universal gravitation constant, and

M_e = mass of the earth.

$$\therefore v = \sqrt{\frac{6.67 \times 10^{-11} \times 5.98 \times 10^{24}}{6.57 \times 10^6}}$$

$$= 7.79 \times 10^3 \text{ m/s.}$$

$$\frac{T^2}{R^3} = \frac{4\pi^2}{GM_e},$$

where T = period of satellite.

$$T^2 = \frac{4\pi^2 R^3}{GM_e}$$

$$\therefore T = \sqrt{\frac{4\pi^2 R^3}{GM_e}}$$

$$= \sqrt{\frac{4\pi^2 (6.57 \times 10^6)^3}{6.67 \times 10^{-11} \times 5.98 \times 10^{24}}}$$

$$= \sqrt{28.069 \times 10^6}$$

$$= 5.3 \times 10^3 \, \text{s}$$

$$= 1.47 \, \text{hr}$$

Example 2

A satellite orbits the earth at the equator with a period of 86,400 seconds.
Given:

Mass of the earth = 5.98×10^{24} kg,

Radius of the earth = 6.37×10^6 m,

Universal gravitation constant = 6.67×10^{-11} Nm^2/kg^2, and

Speed of light in space = 3×10^8 m/s, calculate:
   a) The height of the satellite above the surface of the earth.
   b) The speed of the satellite.

c) The minimum time taken for the signal from the satellite to reach the earth.

Solution

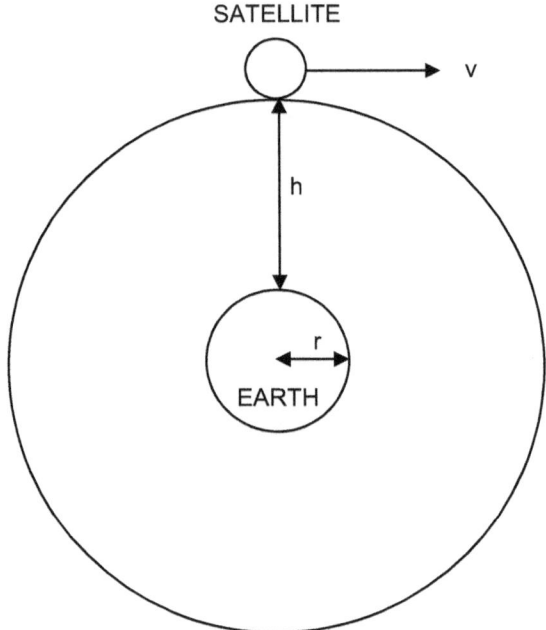

Fig. 6.13: Figure for Example 2

a) $\dfrac{T^2}{R^3} = \dfrac{4\pi^2}{GM_e}.$

$\therefore R^3 = \dfrac{T^2 GM_e}{4\pi^2}.$

$\therefore R = \sqrt[3]{\dfrac{T^2 GM_e}{4\pi^2}}$

$$= \sqrt[3]{\frac{86400^2 \times 6.67 \times 10^{-11} \times 5.98 \times 10^{24}}{4\pi^2}}$$

$$= \sqrt[3]{75.4214 \times 10^{21}}$$

$$= 4.225 \times 10^7 \text{ m}$$

$R = h + r$

$\therefore h = R - r$

$$= 4.225 \times 10^7 - 6.37 \times 10^6$$

$$= 3.588 \times 10^7 \text{ m}.$$

b) $v = \dfrac{2\pi R}{T}$.

$$= \frac{2\pi \times 4.225 \times 10^7}{86400}.$$

$$= 3.07 \times 10^3 \text{ m/s}.$$

c) The minimum distance from the satellite to the earth is $h = 3.588 \times 10^7$ m.

Speed of light $c = 3 \times 10^8$ m/s.

Hence, the minimum time required is given by:

$$t_{min} = \frac{h}{c}$$

$$= \frac{3.588 \times 10^7}{3 \times 10^8}$$

$$= 0.1196 \text{ s}$$

= 120 ms.

Example 3

A satellite orbits the earth with a radius of 16370 km. Given:

Mass of the earth = 5.98×10^{24} kg,

Radius of the earth = 6.37×10^6 m, and

Universal gravitation constant = 6.67×10^{-11} Nm²/kg²,

a) Calculate the speed of the satellite.
b) Calculate the acceleration of the satellite.
c) Calculate the orbital period of the satellite.
d) What is this type of satellite called and how many of them would be required to give global coverage?

Solution

a) Radius of orbit, $R = 16.37 \times 10^6$ m.

The velocity is given by:

$$v = \sqrt{\frac{GM_e}{R}} \text{ m/s},$$

$$\therefore v = \sqrt{\frac{6.67 \times 10^{-11} \times 5.98 \times 10^{24}}{16.37 \times 10^6}}$$

$$= 4.94 \times 10^3 \text{ m/s}.$$

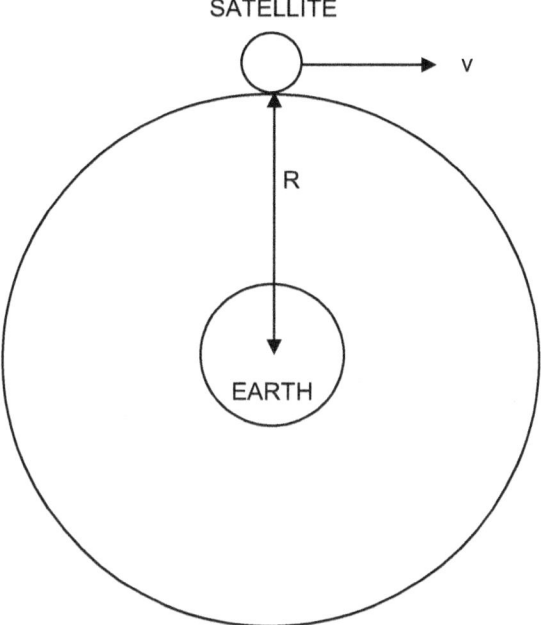

Fig. 6.14: Figure for Example 3

b) The acceleration is given by:

$$a = \frac{GM_e}{R^2}.$$

$$= \frac{6.67 \times 10^{-11} \times 5.98 \times 10^{24}}{(16.37 \times 10^6)^2}.$$

$$= 1.49 \text{ m/s}^2.$$

c) $\dfrac{T^2}{R^3} = \dfrac{4\pi^2}{GM_e}.$

$$\therefore T = \sqrt{\frac{4\pi^2 R^3}{GM_e}}$$

$$= \sqrt{\frac{4\pi^2 (16.37 \times 10^6)^3}{6.67 \times 10^{-11} \times 5.98 \times 10^{24}}}$$

$$= \sqrt{4.343 \times 10^8}$$

$$= 20837 \text{ s}$$

$$= 5.79 \text{ hr}$$

d) This type of satellite is called Medium Earth Orbit Satellite (MEO). 12 MEO satellites are required for global coverage.

6.9 Questions on Satellite Communication

Question 1

The earth orbits the sun with a period of 24 hours. Given the following:

Mass of the sun = 1.99×10^{30} kg,

Radius of the earth = 6.955×10^8 m, and

Universal gravitation constant = 6.67×10^{-11} Nm²/kg², calculate:

a) The height of the earth above the surface of the sun.
b) The speed of the earth.

Answer: 2.23×10^9 m; 2.13×10^5 m/s.

Question 2

A satellite orbits the earth with a speed of 7.62×10^3 m/s. Calculate:

a) The height of the satellite above the earth.

b) The orbital period of the satellite.

Answer: 500 km; 1.57 hrs.

Question 3

A satellite orbits the earth with an acceleration of 1.7 m/s². Calculate:

a) The height of the satellite above the earth.

b) The orbital period of the satellite.

Answer: 8.95×10^6 m; 5.24 hrs.

References

[1] Kennedy et al.: Electronic Communication Systems. McGraw - Hill. International Edition 1992.

[2] M. Connor.: Telecommunication Systems and Services. 1985.

[3] John Crisp.: Introduction to Fiber Optics. Newnes Second Edition 2001.

[4] Behrouz A. Forouzan.: Data Communications and Networking. McGraw - Hill Second Edition 2001.

[5] W. Fraser.: Telecommunication – An Introductory Textbook for Engineering Students. Macdonald 1963.

[6] The Distance Learning Office. Brunel University.: Satellite / Optical and Mobile Communication Systems.Issue 6. Section 2: Satellite Communication Systems 2001.

www.ingramcontent.com/pod-product-compliance
Lightning Source LLC
Chambersburg PA
CBHW020644220526
45464CB00001B/287